Evolution of Herbivory in Terrestrial Vertebrates
Perspectives from the fossil record

Although herbivory probably first appeared over 300 million years ago, it only became established as a common feeding strategy during Late Permian times. Subsequently, herbivory evolved in numerous lineages of terrestrial vertebrates, and the acquisition of this mode of feeding was frequently associated with considerable evolutionary diversification in those lineages. This book represents the first comprehensive overview of the evolution of herbivory in land-dwelling amniote tetrapods in recent years. In *The Evolution of Herbivory in Terrestrial Vertebrates* leading experts review the structural adaptations for, and the evolutionary history of, feeding on plants in the major groups of land-dwelling vertebrates, especially dinosaurs and ungulate mammals. As such it will be the definitive reference source on this topic for evolutionary biologists and vertebrate paleontologists.

HANS-DIETER SUES is Vice-President for Collections and Research at the Royal Ontario Museum in Toronto, and Professor of Zoology at the University of Toronto. He is interested in the evolution of late Paleozoic and Mesozoic tetrapods, especially dinosaurs and their relatives, and patterns of ecosystem change through time. Professor Sues has co-edited two other volumes, *Evolution of Terrestrial Ecosystems Through Time* (1992) and *In the Shadow of the Dinosaurs: Early Mesozoic Tetrapods* (1994).

Evolution of Herbivory in Terrestrial Vertebrates

Perspectives from the fossil record

Edited by

HANS-DIETER SUES

Royal Ontario Museum and University of Toronto

CAMBRIDGE UNIVERSITY PRESS
Cambridge, New York, Melbourne, Madrid, Cape Town, Singapore, São Paulo

Cambridge University Press
The Edinburgh Building, Cambridge CB2 2RU, UK

Published in the United States of America by Cambridge University Press, New York

www.cambridge.org
Information on this title: www.cambridge.org/9780521594493

First published 2000
This digitally printed first paperback version 2005

A catalogue record for this publication is available from the British Library

Library of Congress Cataloguing in Publication data

Evolution of herbivory in terrestrial vertebrates: perspectives from the fossil record /
edited by Hans-Dieter Sues.
p. cm.
Includes index.
ISBN 0 521 59449 9 (hardcover : alk. paper)
1. Herbivores, Fossil–Evolution. 2. Vertebrates, Fossil. I. Sues, Hans-Dieter, 1956–
QE841.E96 2000
566–dc21 00-020863

ISBN-13 978-0-521-59449-3 hardback
ISBN-10 0-521-59449-9 hardback

ISBN-13 978-0-521-02119-7 paperback
ISBN-10 0-521-02119-7 paperback

Contents

Contributors

PAUL M. BARRETT
Department of Zoology, University of Oxford, South Parks Road, Oxford OX1 3PS, UK

CHRISTINE M. JANIS
Department of Ecology and Evolutionary Biology, Brown University, Providence, RI 02912

CORALIA-MARIA JIANU
Muzeul Civilizatiei Dacice si Romane Deva, B-dul 1 Decembrie Nr. 39, 2700 Deva, Romania

BRUCE J. MACFADDEN
Florida Museum of Natural History, University of Florida, Gainesville, FL 32611

ROBERT R. REISZ
Department of Biology, Erindale College, University of Toronto, 3359 Mississauga Road, Mississauga, ON L5L 1C6, Canada

JOHN M. RENSBERGER
Burke Museum and Department of Geological Sciences, Box 353010, University of Washington, Seattle, WA 98195

HANS-DIETER SUES
Royal Ontario Museum, 100 Queen's Park, Toronto, ON M5S 2C6, and Department of Zoology, University of Toronto, Toronto, ON M5S 1A1, Canada

PAUL UPCHURCH
School of Biological Sciences, University of Bristol, Woodland Road, Bristol BS8 1UG, UK

DAVID B. WEISHAMPEL
Department of Cell Biology and Anatomy, The Johns Hopkins University School of Medicine, 725 North Wolfe Street, Baltimore, MD 21205

Preface

This book presents a collection of paleontological perspectives on the origin and evolution of herbivory in various major taxa of terrestrial vertebrates. The complex interactions between plants and their animal consumers have long been the subject of much interest to evolutionary biologists. At the same time, most studies exploring coevolutionary relationships have focused on extant organisms, and generally little attention has been paid to the historical development of plant–animal interactions through time documented by the fossil record.

Most contributors to this volume review the nature and acquisition of structural features of the skull and dentition suitable for feeding on high-fiber plant material in various major lineages of herbivorous tetrapods. In some instances, they also discuss other lines of evidence (such as isotopic data) bearing on this issue as well as the possible impact of herbivory on the evolutionary diversification of that group. Traditionally, paleobiological studies have assumed a direct relationship between form and function, but current research on the functional morphology of extinct organisms is much more mindful of the inherent theoretical and practical difficulties in reconstructing the habits of ancient organisms. The present volume cannot and does not provide a comprehensive account of herbivory in extinct vertebrates. Rather, it is intended as a review of current research on some of the key issues for advanced students of evolutionary biology, historical ecology, and paleobiology and, it is to be hoped, as a stimulus for further work.

Most chapters are based on contributions presented at a symposium on the evolution of herbivory in insects and terrestrial vertebrates held during the Sixth North American Paleontological Convention (NAPC-96) at the Smithsonian Institution in Washington, D.C. in June 1996. Special thanks are due to the co-organizer of this symposium, Conrad C. Labandeira (National Museum of Natural History), for his co-operation and for sharing my enthusiasm for this subject, and to the Organizing Committee of NAPC-96 for its interest and support.

As editor, I am indebted first to the contributors, for taking part in this project, for (in some cases) meeting deadlines, and for cheerfully putting up with my editorial efforts. At Cambridge University Press, I would like to thank Robin Smith, who first encouraged me to compile this collection of papers, and Tracey Sanderson for her continuing encouragement and support. During the editing of this volume, I have heavily relied on the expertise of many reviewers. I would like to acknowledge the generous assistance and thoughtful comments provided by Richard Beerbower (State University of New York, Binghamton), Robert L. Carroll (McGill University), Peter Dodson (University of Pennsylvania), Robert J. Emry (Smithsonian Institution), James O. Farlow (Indiana University–Purdue University, Fort Wayne), Mikael Fortelius (University of Helsinki), the late Nicholas Hotton III (Smithsonian Institution), John P. Hunter (New York College of Osteopathic Medicine, Old Westbury), Christine M. Janis (Brown University), Gillian King (University of Cambridge), Paul L. Koch (University of California at Santa Cruz), David B. Norman (University of Cambridge), Robert R. Reisz (Erindale College, University of Toronto, Mississauga), Richard K. Stucky (Denver Museum of Natural History), David B. Weishampel (The Johns Hopkins University), and Jeffrey A. Wilson (University of Chicago). I thank Janice Robertson for her careful copy-editing and especially Joan Burke (Royal Ontario Museum) for her meticulous proof-reading.

Hans-Dieter Sues

HANS-DIETER SUES

1

Herbivory in terrestrial vertebrates: an introduction

Introduction

Understanding the ecological attributes of extinct organisms has long been a major research topic in paleobiology, dating back to the pioneering work of the French paleontologist Georges Cuvier in the early nineteenth century. Inferences concerning the ecology of an extinct organism can be based on functional interpretation of its structure, by analogy with present-day relatives, or from the sedimentary context and distribution of fossil remains referable to this taxon (Wing *et al.* 1992). Traditionally, functional morphology has been the most widely used of these approaches. It basically relies on the analysis of organisms as simple machines with functional attributes that can be inferred from the physical properties of their bodies as well as from their shape and size. Chemical analyses of hard tissues (such as extraction of preserved stable carbon isotopes) increasingly are providing significant new data for inferring diet in extinct animals. In recent years, researchers have developed various procedures for linking inferences concerning function in fossils to phylogenetic analyses, increasing confidence in the robustness of these reconstructions (see various papers in Thomason [1995]).

Herbivory, the consumption of plant tissues, is a widespread phenomenon among terrestrial vertebrates. It has frequently and independently evolved in many lineages of amniotes during the last 300 million years or so. Some major groups of herbivorous tetrapods, such as ungulate mammals and ornithischian dinosaurs, attained great abundance and taxonomic diversity. Indeed, the advent of herbivory among land-dwelling tetrapods was one of the key events in the history of life on land. It led to the establishment of 'modern' continental ecosystems, with vast numbers of herbivores supporting a relatively small number of carnivores, during the Permian period (Olson 1966; Hotton *et al.* 1997; Sues and Reisz 1998).

Feeding on plants requires many morphological and physiological modifications to facilitate the efficient reduction and digestion of plant tissues. Plant fodder contains less caloric energy per volume unit than do animal foods (Southwood 1973). Furthermore, in most instances, much of that energy is tied up in substances that are difficult to digest for vertebrates. The contents of plant cells are enclosed by walls that are primarily composed of cellulose, hemicellulose, and lignin. Cellulose is a polymer of glucose, but its glucose units are linked together in such a fashion that it cannot be readily broken down. This makes plant tissues, to varying degrees, more difficult to digest than animal tissues, which are devoid of resistant cell walls. Extant vertebrates lack any endogenous enzymes to hydrolyze the compounds forming the cell walls of plant tissues, but many micro-organisms (bacteria, protists) can produce them. Thus many plant-eating tetrapods have entered into endosymbiotic relationships with such organisms to facilitate cellulysis in their digestive tracts, resulting in the production of sugars and volatile fatty acids that can be absorbed by the vertebrate host. Not only are the walls of the plant cells themselves resistant to unaided digestion by vertebrates but they often protect the digestible cellular contents such as lipids, sugars, organic acids and proteins. Initial breakdown of the cell walls by mechanical or chemical action is thus required.

Plant tissues are highly variable in terms of their nutritional value to vertebrate consumers. Certain plant parts, such as fruits, seeds, and immature vegetative tissues, contain much digestible matter that is only protected by relatively delicate cell walls and often require only little processing to make the enclosed nutrients available to the consumer. However, these are seasonally available resources, and most herbivores subsist on the tougher, more cellulose-rich vegetative structures, such as leaves, roots, shoots and stems that form the bulk of available plant material. The effective utilization of such fodder by the vertebrate consumer requires two steps. First, the plant material must be mechanically broken up by oral processing or by comminution in a muscular foregut or gizzard using ingested grit and pebbles. Grinding the plant fodder more finely clearly increases the rate at which herbivores can process it (Bjorndal *et al.* 1990). Second, symbiotic micro-organisms in the gut must convert the cellulose into volatile fatty acids (especially acetic, butyric, and proprionic acid) that can be readily absorbed by the vertebrate host. Not all extant herbivorous vertebrates employ microbial endosymbionts. Some animals such as the giant panda *Ailuropoda melanoleuca* (Schaller *et al.* 1985) apparently compensate for this lack by consuming large amounts of plant fodder as well as maintaining relatively low activity and growth rates (see McNab 1986).

Most studies of herbivory in present-day vertebrates have focused on the digestive performance of plant-eating mammals, particularly ungulates, due to the commercial importance of the latter (McBee 1977; Chivers and Langer 1994). The adaptations for feeding on plants in non-mammalian tetrapods are much less well studied. Relatively few taxa of present-day reptiles are obligate herbivores, and the range of structural features of the skull and dentition for feeding on plants is limited compared with that in mammals (Throckmorton 1976). However, more recent work has demonstrated that plant-eating iguanid lizards can degrade cellulose and hemicellulose almost as efficiently as herbivorous mammals (Troyer 1984).

Herbivory in extinct vertebrates

Unequivocal evidence for trophic interactions between extinct animals and plants is only rarely found. Therefore, inferences concerning herbivory in extinct vertebrates must rely almost entirely on circumstantial evidence. Starting with Cuvier's pioneering work, paleontologists have correlated broadly defined feeding categories with specific morphological attributes, primarily of the skull and dentition, in tetrapods. Such correlations are based either on analogy to present-day animals with known dietary habits or on biomechanical assessment of the suitability of tooth structure for processing potential food items (e.g., Rensberger 1973; Lucas 1979). In a few instances, fossilized gut contents provide a more direct line of evidence, but it is important to rule out postmortem introduction of material into the abdominal cavity. Furthermore, some food stuffs may have already been digested prior to the death and fossilization of the consumer.

Mammals are distinguished from most other tetrapods by extensive oral processing of food or mastication. Thus their masticatory apparatus provides an excellent model for correlating tooth structure with diet. The specific relationships between gross occlusal structures and mechanical performance can be modeled (Rensberger 1973; Lucas 1979; Lucas and

Luke 1984) and, in many instances, can be tested experimentally in extant forms (Kay and Sheine 1979; Teaford and Oyen 1989). Wear on tooth enamel at the microscopic level can be related to diet and the masticatory system (Walker *et al.* 1978; Rensberger 1986), and complex functional associations exist between enamel microstructure, gross occlusal features, and chewing direction (Koenigswald 1980; Rensberger and Koenigswald 1980). As teeth are the most commonly preserved part of the mammalian skeleton, they thus become a valuable resource for paleontologists and physical anthropologists concerned with paleoecological questions.

In non-mammalian vertebrates, the correlation between diet and dentition is much more difficult because oral processing of food and thus tooth occlusion are uncommon. Furthermore, in some groups (especially birds) with herbivorous taxa, teeth are absent altogether. However, based on biomechanical considerations and comparison with extant forms, it is still possible in many cases to interpret craniodental attributes in general terms of their biomechanical suitability for feeding on a particular type of material.

Dental features suggestive or indicative of high-fiber herbivory include dentitions adapted for crushing and grinding, or marginal teeth with labiolingually compressed, leaf-shaped, and cuspidate crowns suitable for puncturing and shredding plant fodder. Features of the skull and mandible variously associated with feeding on plants, especially in mammals, include short tooth rows (along with foreshortening of the snout and mandible), elevation/depression of the jaw joint relative to the occlusal plane for increased mechanical advantage of the adductor jaw muscles, enlargement of the adductor chambers and temporal openings as well as deepening of the zygomatic arches and mandibular rami for the origin and insertion of substantial adductor jaw muscles, and jaw joints suitable for complex mandibular motion (Maynard Smith and Savage 1959; Turnbull 1970; Rensberger 1986; Janis 1995). Finally, plant-eating tetrapods typically have longer and/or bulkier digestive tracts, and a longer and/or broader trunk region, than related faunivorous forms (Schiek and Millar 1985; Dearing 1993) because part of the gut is modified to form a reservoir housing and creating suitable pH conditions for the endosymbiotic micro-organisms involved in the fermentative breakdown of the cellulose from the ingested plant fodder. Passage of food through longer intestinal tracts also allows for longer periods of time for processing of resistant materials. Thus the rib cages of many herbivorous tetrapods are either much wider and more capacious ('barrel-shaped') than those of their closest faunivorous relatives, or the trunk region is elongated to accommodate a longer digestive tract. This is reflected in overall body proportions as well as in the structure of the vertebral column.

In this volume, nine contributors examine the structural correlates of herbivory in various taxa of herbivorous tetrapods and discuss them in relation to the evolutionary diversification of these groups.

Reisz and Sues (Chapter 2) review the earliest occurrences of herbivory in amniote tetrapods. They use the craniodental and, to a lesser extent, postcranial features discussed above to identify probably herbivorous taxa among late Paleozoic and Triassic amniotes. The oldest known putative herbivores, of Late Pennsylvanian age, are the Diadectidae, a group of tetrapods very closely related to the Amniota, and the basal synapsid *Edaphosaurus*. Surprisingly, feeding on plants only became common among tetrapods during the Late Permian. At that point in time, vertebrate herbivores first became a key component of the terrestrial food web, leading to the establishment of the basic trophic structure of modern terrestrial ecosystems (Olson 1961, 1966). Phylogenetic analysis indicates that herbivory was repeatedly and independently acquired in various lineages of amniotes during the Late Carboniferous and Permian (Sues and Reisz 1998).

In his discussion of possible diets for prosauropod dinosaurs, Barrett (Chapter 3) injects a cautionary note concerning the identification of herbivory in non-mammalian tetrapods based primarily on dental features. Citing anecdotal evidence, he argues that present-day iguanid lizards, which have teeth with leaf-shaped, serrated crowns, have long provided a model for inferring herbivory in certain dinosaurs with similar teeth, but are actually omnivorous rather than strictly herbivorous. However, few extant vertebrates are exclusively herbivorous, and many herbivores supplement their diet by the intake of animal protein.

Upchurch and Barrett (Chapter 4) review the craniodental and postcranial features of the constituent clades of sauropod dinosaurs in relation to possible feeding strategies. They marshal evidence ranging from patterns to dental microwear to browsing height inferred from skeletal proportions to demonstrate a diversity of possible feeding styles among sauropods. Sauropodomorph dinosaurs (prosauropods and sauropods) are noteworthy because they represented the first diverse radiation of megaherbivores (*sensu* Owen-Smith [1988], with body weights exceeding one metric ton) and were capable of foraging at greater heights (i.e., more than one or two meters above the ground) than other plant-eating tetrapods at that time.

Weishampel and Jianu (Chapter 5) present a detailed phylogenetic analysis of the various major taxa of dinosaurian herbivores. They emphasize the importance of taking into consideration the unrecorded segments of lineages ('ghost lineages') that can be inferred based on phylogenetic hypotheses in estimating overall diversity. The authors find little evidence to support the currently popular coevolutionary scenario linking the onset of the evolutionary diversification of flowering plants (angiosperms) to the radiations of large ornithopod and ceratopsian dinosaurs.

Rensberger (Chapter 6) provides an elegant analysis of the biomechanical factors dictating tooth configuration in herbivorous placental mammals from the early Cenozoic. In the two most common groups of early Paleocene ungulates in North America, the shearing component of mastication was greatly reduced relative to the condition in more primitive mammals and compression became the dominant component. Rensberger demonstrates that the stresses in the tooth enamel induced by chewing are lower in low, wide cusps than in tall, sharp cusps. Prism decussation (where zones of prisms with a common orientation alternate with zones of prisms with a different orientation) in the enamel increases resistance to fracturing. It appears in most ungulates as body size increases later during the Paleogene. With the acquisition of stronger enamel, the earlier trend toward more blunt cusps is reversed, and shearing crests reappear.

Based on a recent compendium of Tertiary mammals from North America, Janis (Chapter 7) reviews the diversification of the different types of feeding strategies (as deduced from tooth shape) in herbivorous mammals from the Paleogene (Paleocene–Oligocene) of North America. She relates the relative abundance of the different kinds of tooth shape to climatic changes during the Early Tertiary inferred from other lines of evidence. The early 'condylarths' were presumably omnivorous rather than strictly herbivorous. Feeding on foliage (folivory) apparently did not occur until the latest Paleocene. The earliest artiodactyls and perissodactyls appear in the early Eocene, but they had dentitions more typical of omnivores/frugivores, and folivory in these groups was not common until the late Eocene. Although ungulates with relatively high-crowned teeth are known from the Paleogene, there is no evidence of true grazers in the fossil record until the Neogene.

MacFadden (Chapter 8) presents an overview of recent work by him and various colleagues, including the application of isotopic analyses, on the origin and early evolution of the grazing guild among Neogene mammals from the Americas. Grazing first developed as a major feeding strategy during the Miocene. The acquisition of high-crowned teeth, which appears to be an adaptation to the pervasive occurrence of highly abrasive silica particles in grasses, is interpreted as a coevolutionary response to the advent of widespread grassland communities at that time. Grazing was independently acquired in various North American groups (artiodactyls, perissodactyls, proboscideans, rodents) and in the South American notoungulates prior to the formation of the Panamanian land bridge. Based on MacFadden's work, grasslands and grazing mammals may have originated slightly earlier in South America than in North America. Data for stable carbon isotopes indicate that the earliest grasslands were formed by C_3 grasses. The change to the grasslands dominated by C_4 grasses (which comprise most grasslands today) occurred after the late Miocene and may explain the observed decline in grazing diversity after the late Miocene.

References

Bjorndal, K. A., Bolten, A. B., and Moore, J. E. (1990). Digestive fermentation in herbivores: effect of food particle size. *Physiol. Zool.* 63:710–721.

Chivers, D. J., and Langer, P. (eds.) (1994). *The Digestive System in Mammals: Food, Form and Function.* Cambridge and New York: Cambridge University Press.

Dearing, M. D. (1993). An alimentary specialization for herbivory in the tropical whiptail lizard *Cnemidophorus murinus*. *J. Herpet.* 27:111–114.

Hotton, N. III, Olson, E.C., and Beerbower, R. (1997). Amniote origins and the discovery of herbivory. In *Amniote Origins*, ed. S. S. Sumida and K. L. M. Martin, pp. 207–264. San Diego: Academic Press.

Janis, C. M. (1995). Correlations between craniodental morphology and feeding behavior in ungulates: reciprocal illumination between living and fossil taxa. In *Functional Morphology in Vertebrate Paleontology*, ed. J. J. Thomason, pp.76–98. Cambridge and New York: Cambridge University Press.

Kay, R. F., and Sheine, W. S. (1979). On the relationship between chitin particle size and digestibility in the primate *Galago senegalensis*. Amer. J. Phys. Anthrop. 50:301–308.

Koenigswald, W. von. (1980). Schmelzstruktur und Morphologie in den Molaren der Arvicolidae (Rodentia). *Abh. Senckenberg. Naturforsch. Ges.* 539:1–129.

Lucas, P. W. (1979). The dental-dietary adaptations of mammals. *N. Jb. Geol. Paläont., Mh.* 1979:486–516.

Lucas, P. W., and Luke, D. A. (1984). Chewing it over: basic principles of food breakdown. In *Food Acquisition and Processing in Primates*, ed. D. A. Chivers, B. A. Wood, and A. Bilsborough, pp. 283–301. New York: Plenum Press.

Maynard Smith, J., and Savage, R. J. G. (1959). The mechanics of mammalian jaws. *School Sci. Rev.* 141:289–301.
McBee, R. H. (1977). Fermentation in the hindgut. In *Microbial Ecology of the Gut*, ed. R. T. J. Clarke and T. Bauchop, pp. 185–222. London and New York: Academic Press.
McNab, B. K. (1986). The influence of food habits on the energetics of eutherian mammals. *Ecol. Monographs* 56:1–19.
Olson, E. C. (1961). The food chain and the origin of mammals. In *International Colloquium on the Evolution of Lower and Non-Specialized Mammals*, ed. G. Vanderbroek, pp. 97–116. Brussels: Koninklijke Vlaamse Akademie voor Wetenschappen, Letteren en Schone Kunsten van België.
Olson, E. C. (1966). Community evolution and the origin of mammals. *Ecology* 47:291–308.
Owen-Smith, R. N. (1988). *Megaherbivores: The Influence of Very Large Body Size on Ecology*. Cambridge and New York: Cambridge University Press.
Rensberger, J. M. (1973). An occlusal model for mastication and dental wear in herbivorous mammals. *J. Paleont.* 47:515–528.
Rensberger, J. M. (1986). Early chewing mechanisms in mammalian herbivores. *Paleobiology* 12:474–494.
Rensberger, J. M., and Koenigswald, W. von. (1980). Functional and phylogenetic interpretation of enamel microstructure in rhinoceroses. *Paleobiology* 6:477–495.
Schaller, G. B., Hu J., Pan W., and Zhu J. (1985). *The Giant Pandas of Wolong*. Chicago: University of Chicago Press.
Schiek, J. O., and Millar, J. S. (1985). Alimentary tract measurements as indicators of diet in small mammals. *Mammalia* 49:93–104.
Southwood, T. R. E. (1973). The insect/plant relationship – an evolutionary perspective. *Symp. Roy. Soc. Lond.* 6:3–30.
Sues, H.-D., and Reisz, R. R. (1998). Origin and early evolution of herbivory in terrestrial tetrapods. *Trends Ecol. Evol.* 18:141–145.
Teaford, M. F., and Oyen, O. J. (1989). Differences in the rate of molar wear between monkeys raised on different diets. *J. Dental Res.* 68:1513–1518.
Thomason, J. J. (ed.) (1995). *Functional Morphology in Vertebrate Paleontology*. Cambridge and New York: Cambridge University Press.
Throckmorton, G. S. (1976). Oral food processing in two herbivorous lizards, *Iguana iguana* (Iguanidae) and *Uromastix aegyptius* (Agamidae). *J. Morph.* 48:363–390.
Troyer, K. (1984). Structure and function of the digestive tract of a herbivorous lizard, *Iguana iguana*. *Physiol. Zool.* 57:1–8.
Turnbull, W. D. (1970). Mammalian masticatory apparatus. *Fieldiana Geol.* 18:149–356.
Walker, A., Hoeck, N., and Perez, L. (1978). Microwear of mammalian teeth as an indicator of diet. *Science* 201:908–910.
Wing, S. L., Sues, H.-D., Potts, R., DiMichele, W. A., and Behrensmeyer, A. K. (1992). Evolutionary paleoecology. In *Terrestrial Ecosystems through Time*, ed. A. K. Behrensmeyer *et al.*, pp. 1–13. Chicago: University of Chicago Press.

ROBERT R. REISZ AND HANS-DIETER SUES

2

Herbivory in late Paleozoic and Triassic terrestrial vertebrates

Introduction

The exploitation of land plants as a major food resource by amniote tetrapods led to profound changes in the pattern of trophic interactions in terrestrial ecosystems during the late Paleozoic (Olson 1961, 1966; King 1996; Hotton *et al.* 1997; Sues and Reisz 1998). Prior to the appearance of various forms specialized for feeding on plants, tetrapods could access the plentiful vegetal resources only indirectly through detritivory and consumption of invertebrates that fed on plants and/or plant detritus (Olson 1961, 1966). The oldest plant-eating tetrapods are known from the Late Pennsylvanian (Late Carboniferous) of North America and Europe. However, herbivores did not form a major component of the known terrestrial tetrapod assemblages from the Early Permian. Only during the Late Permian, some 40 million years after their first appearance, did plant-eating tetrapods become abundant and much diversified in the fossil record. At that time, a 'modern' pattern of trophic interactions was established, with a vast standing crop of herbivores sustaining a relatively small number of top carnivores. Communities of this type are first documented by diverse assemblages of tetrapods from the Late Permian portion of the Beaufort Group of South Africa (Kitching 1977) and, to a lesser extent, from more or less coeval continental strata in other regions of the world.

Morphological and physiological correlates of herbivory

The effective utilization of high-fiber plant material by vertebrates requires two steps. First, the plant material must be mechanically broken up by oral processing or by comminution in a muscular gizzard using ingested grit and pebbles. Present-day herbivorous lizards do not chew their food (Throckmorton 1976), and leaves often pass through the gastrointestinal tract virtually intact (Bjorndal 1979; Iverson 1982). Grinding the plant fodder more finely clearly increases the rate at which herbivores can process it (Bjorndal *et al.* 1990). Second, symbiotic micro-organisms in the gut must convert the cellulose into volatile fatty acids (especially acetic, butyric, and proprionic acid) that can be readily absorbed by the vertebrate host.

In this chapter, we review the possible exploitation of high-fiber plant material by various groups of late Paleozoic and Triassic terrestrial tetrapods. We discuss the earliest stages in the evolution of herbivory among terrestrial tetrapods and the possible impact of this feeding strategy on the evolution of continental vertebrate communities. It is important to stress here that relatively few extant reptiles are obligate herbivores (see also Barrett, this volume), and most feed on a variety of dietary items (see below). By analogy, we expect similar dietary versatility among late Paleozoic and early Mesozoic amniotes. However, we should note that the skeletal modifications observed in certain extinct forms are, among extant vertebrates, most commonly associated with diets that include high-fiber plant material as the primary component.

In most cases, inferences concerning the dietary habits of extinct vertebrates rely on circumstantial evidence. Dental features suggestive or indicative of high-fiber herbivory include dentitions adapted for crushing and grinding, or marginal teeth with labiolingually compressed, leaf-shaped, and cuspidate crowns suitable for puncturing and shredding plant fodder. Cranial features variously associated with feeding on plants, especially in mammals, include short tooth rows (and related foreshortening of the snout and mandible), elevation or depression of the jaw joint relative to the occlusal plane for increased mechanical advantage of the adductor jaw muscles during biting, enlargement of the adductor chambers and temporal openings as well as deepening of the zygomatic arches and mandibular rami to accommodate powerful adductor jaw muscles, and jaw joints suitable for complex mandibular movements (Maynard Smith and Savage 1959; Turnbull 1970; Rensberger 1986; Janis 1995). Herbivorous tetrapods typically have longer and/or bulkier digestive tracts, and a longer and/or broader trunk region, than related faunivorous forms (Schiek and Millar 1985; Dearing 1993) because part of the gut is modified to form a reservoir housing and creating suitable pH conditions for endosymbiotic micro-organisms for fermentative breakdown of the cellulose from the ingested plant fodder. Passage of food through longer intestinal tracts allows for longer periods of time for the processing of resistant materials. Thus the rib cages of most early herbivorous tetrapods (especially caseid synapsids) are either much wider and more capacious ('barrel-shaped') than those of their closest faunivorous relatives, or, alternatively, the trunk region is distinctly elongated (e.g., in many dicynodonts) to accommodate a longer digestive tract. These features are reflected in overall body proportions and frequently in the structure of the vertebral column as well.

The underlying assumption of this review is that high-fiber plant parts formed the primary component of the diet for an extinct tetrapod if the dental and skeletal features of this animal together are structurally suitable for processing of fiber-rich foodstuffs. Once herbivory has been established with some degree of confidence in an individual taxon or a group of closely related taxa, we can examine the anatomical correlates in an explicitly phylogenetic context to trace their historical development. Although the reasoning behind this comparative approach is admittedly somewhat circular, it permits relating the development of certain features of the feeding apparatus to the adoption of a herbivorous diet.

Acquisition of cellulytic endosymbionts by herbivorous tetrapods

The aforementioned anatomical adaptations for feeding on high-fiber plant material would be relatively ineffectual in the absence of endosymbiotic micro-organisms for cellulysis. The actual mode of acquisition of these endosymbionts by their vertebrate hosts is unknown, and it is unlikely that the fossil record will ever provide direct clues for resolving this issue. Hotton *et al.* (1997) suggested that the micro-organisms may have initially been picked up by detritivorous animals foraging in plant litter. Those that could survive in the tetrapod gut could have gradually assumed a role in the digestive processes of the host. We tested this hypothesis by considering the phylogenetic relationships of each major lineage of presumed late Paleozoic and Triassic herbivorous tetrapods. We found that, in most cases, their sister-taxa were insectivorous, and thus we hypothesize that most of these herbivorous forms were derived from insect-consuming precursors. Gow (1978) argued the same point for the various groups of plant-eating reptiles and non-mammalian synapsids considered in his brief review. In some taxa, insectivory may have even been retained in juvenile individuals, followed by a shift toward herbivory in adults (Gow 1978; DeMar and Bolt 1981). Insects were feeding on plants by mid-Carboniferous times, much earlier than tetrapods (Scott and Taylor 1983; Labandeira 1997). We consider it equally plausible that ingested insects, especially those feeding on plant material, provided the original source for fermentative endosymbionts. The currently accepted scheme of tetrapod interrelationships indicates that cellulytic endosymbionts were independently acquired in each lineage (Hotton *et al.* 1997; Sues and Reisz 1998) because the major groups of late Paleozoic plant-eating tetrapods are only distantly related to each other.

High-fiber herbivory in tetrapods presumably developed only after a (possibly extended) transitional period of omnivory that included consumption of plant material because various changes were required for the efficient processing and digestion of plant fodder. The morphological and physiological requirements for omnivory and low-fiber herbivory would not differ significantly from those for strict faunivory (Hotton *et al.* 1997). The present-day herbivorous lizard *Cnemidophorus murinus* is an opportunistic herbivore; indeed, in captivity, this reptile prefers animal foods over any plant material offered (Dearing 1993). Comparative studies on turtles (Bjorndal and Bolten 1993) indicate that herbivores may show greater digestive efficiency than omnivores only on those plant diets that are subject to extensive fermentation; the digestive performance of omnivores may equal, if not exceed, that of herbivores feeding on material that does not require a significant amount of fermentative processing.

Among extant plant-eating iguanid lizards, the juveniles of each generation acquire the requisite microbes for endosymbiotic fermentation by consuming the droppings of adult conspecifics (Troyer 1982). Modesto (1992) argued that this intergenerational type of endosymbiont acquisition could only occur in nest-building amniote tetrapods. Although, at first glance, this hypothesis is attractive because it draws on the most obvious distinction between amniotes and non-amniotes, Modesto's reasoning is not compelling because amphibian hatchlings and juveniles could have easily picked up the endosymbiotic microbes by feeding at or near sites where adults defecated. Most adult extant amphibians are faunivorous, and only one, the Indian green frog (*Rana hexadactyla*), is a folivore (Das 1996).The apparent inability of present-day amphibians to acquire, or at least to maintain, cellulytic endosymbionts is of considerable interest, but remains yet to be explained. In this context, the Permo-

Carboniferous Diadectidae present an interesting conundrum. They presumably were herbivores capable of processing high-fiber plant fodder, but they are not considered amniotes by many recent authors based on osteological features (Reisz 1997). As the closest known relatives of amniotes, diadectids may have been reproductively amniotic, but, in the absence of preserved eggs, their skeletal features place them outside the crown-group Amniota. Perhaps diadectids were not amniotic in reproductive terms, but, being closely related to amniotes, shared with the latter certain anatomical and/or physiological attributes that allowed them to feed on plants (see Hotton *et al.* 1997).

Diversity of late Paleozoic and Triassic herbivorous tetrapods

The known fossil record of amniote tetrapods begins in the Mid-Pennsylvanian (Westphalian B [Moscovian], Joggins, Nova Scotia/Canada; about 308 million years ago [Harland *et al.* 1990]). At that time, the two principal lineages of Amniota, Reptilia (including birds) and Synapsida (including mammals), had already differentiated (Reisz 1997). The first plant-eating amniotes are known from the Late Pennsylvanian (Stephanian B; about 303 to 295 million years ago [Harland *et al.* 1990]). The proximate sister-group of Amniota, Diadectomorpha, which includes presumably herbivorous forms as well (Hotton *et al.* 1997; Sues and Reisz 1998), also first appears in the fossil record during the Late Pennsylvanian. Herbivory is clearly an ancient feeding strategy among amniote tetrapods.

Late Paleozoic and early Mesozoic tetrapods are separated from their present-day relatives by a long evolutionary history, and it is difficult to compare the craniodental features of most of these animals directly with those in their extant relatives. Several fairly robust hypotheses for the interrelationships of the principal groups of late Paleozoic tetrapods have been published in recent years (Laurin and Reisz 1995; Lee 1997; Reisz 1997), and it is now possible to compare the craniodental features of presumed herbivores with the homologous ones in their faunivorous sister-taxa within a phylogenetic framework. This approach permits the identification of potential anatomical features suitable for feeding on plants. For example, the shapes of the tooth crowns in most late Paleozoic tetrapods, including those of the sister-taxa of early herbivores, range from acutely conical to blade-like and recurved with distinct cutting edges. By analogy to most present-day vertebrates with such teeth, this

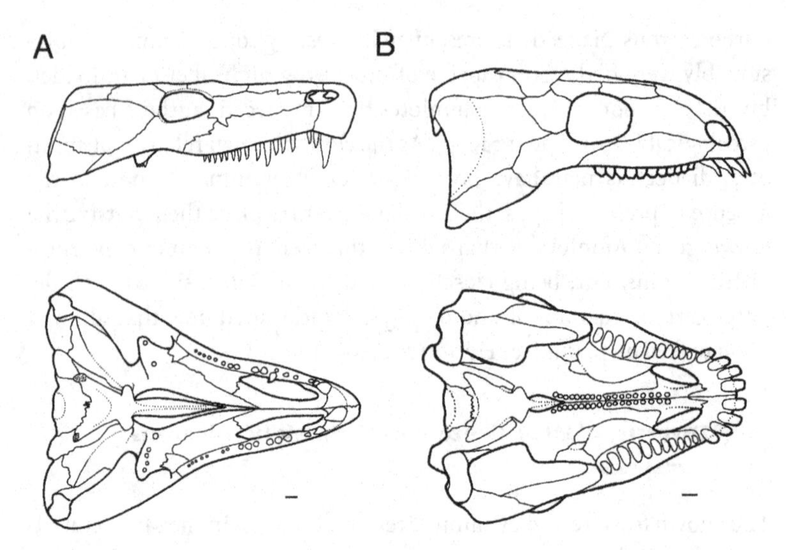

Figure 2.1. Skulls of (A) *Limnoscelis* and (B) *Diadectes* in lateral and palatal views. Scale bars each equal 1 cm. (*Limnoscelis* modified from Williston 1911; *Diadectes* modified from Case 1911 and Olson 1947.)

type of tooth crown was presumably used for immobilizing animal prey and for tearing and slicing through flesh. Only a few Permo-Carboniferous tetrapods had teeth that deviated from this common pattern. The earliest forms with dentitions suitable for processing high-fiber plant fodder date back to the Late Carboniferous, and these animals also have cranial and postcranial features that are consistent with herbivory. Although the fossil record demostrates that feeding on high-fiber plant material had evolved by Late Carboniferous times, it is likely that consumption of other plant tissues already occurred at an earlier date (Hotton *et al.* 1997; Sues and Reisz 1998).

Late Paleozoic taxa

Diadectomorpha: Diadectidae

Desmatodon, from the Late Pennsylvanian of North America, and a closely related but more derived form, *Diadectes* (Figure 2.1B), from the Late Pennsylvanian and Early Permian of North America and Europe, were apparently the earliest herbivores capable of feeding on high-fiber plant material (Hotton *et al.* 1997). Although not as diverse as the older taxonomic literature (e.g., Case 1911) would imply, the fossil record indicates that diadectids formed a common element of the terrestrial tetrapod assemblages of North America and Europe during the Late Pennsylvanian and Early Permian. Most authors now consider Diadectomorpha the proximate sister-taxon of Amniota (Laurin and Reisz 1995; Reisz 1997). However, Lee and Spencer (1997) have recently revived an earlier interpretation of these tetrapods as basal amniotes. We do not find the arguments advanced by these authors persuasive, mainly because their proposal centers on issues of taxonomic stability, rather than phylogenetic relationships. In any case, the clade Diadectomorpha would retain its position outside the crown-group of Amniota.

The spatulate, incisiform anterior teeth of *Diadectes* (Figure 2.1B) are distinctly procumbent. The lower incisiforms occlude with the upper ones; this arrangement is suitable for seizing and cropping food items (Case 1911), which is consistent with the vertical striations produced by wear on the incisiform teeth (Hotton *et al.* 1997). The more posterior dentary and the maxillary teeth of *Diadectes* are molariform, with transversely expanded but anteroposteriorly short crowns suitable for crushing. Longitudinally oriented striations on the typically heavily worn molariform teeth indicate fore-and-aft motion of the mandible (Hotton *et al.* 1997). Berman *et al.* (1998) have found evidence that the lower molariform teeth may have occluded against a pad of perhaps keratinized tissue supported by the palatal shelf of the palatine and the upper molariform teeth against a similar pad on the lingual surface of the parapet of the dentary. *Diadectes* has a large, barrel-shaped trunk that could have accommodated a voluminous digestive tract.

Comparisons with the sister-taxa of Diadectidae, *Limnoscelis* (Figure 2.1A) and *Tseajaia* (Heaton 1980) indicate that these diadectomorphs show the plesiomorphic character-states for the aforementioned features. In addition, there exists a suite of other skeletal features that appear to be related to herbivory. The temporal region of the skull of *Diadectes* is deeper than in other diadectomorphs and basal amniotes, the lower jaw is more massive, and an incipient secondary bony palate is developed. The postcranial skeleton of *Diadectes* is considerably more robust than in other diadectomorphs and basal amniotes, especially in the construction of the trunk region. In particular, the dorsal neural arches of *Diadectes* are tall and massive, with massive neural spines and additional intervertebral articular facets. The large transverse processes supported the strongly curved ribs forming a barrel-shaped trunk (Case 1911). The trunk is proportionately shorter than in other Diadectomorpha; it comprises fewer

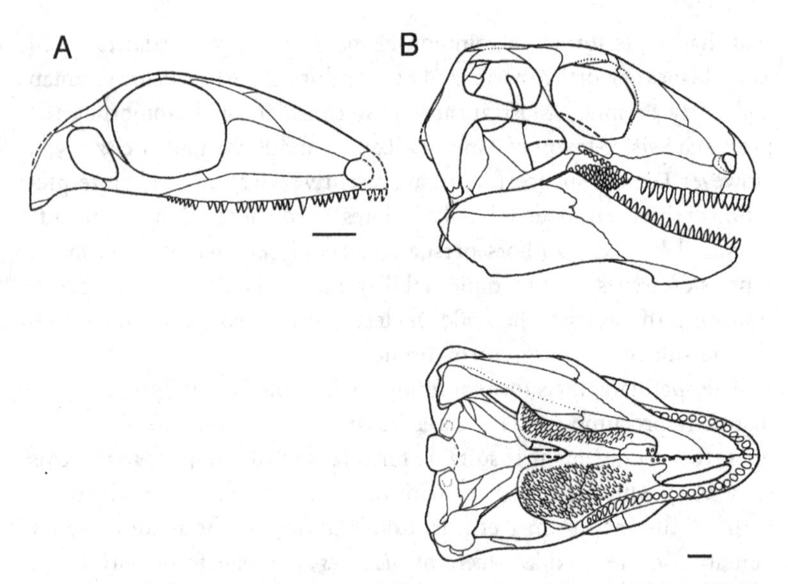

Figure 2.2. Skulls of (A) *Ianthasaurus* and (B) *Edaphosaurus* (with attached mandible) in lateral views and skull and right mandibular ramus of *Edaphosaurus* in palatal view. The palate of *Ianthasaurus* is insufficiently known for a reconstruction. Scale bars each equal 1 cm. (*Ianthasaurus* modified from Modesto and Reisz 1990; *Edaphosaurus* redrawn from Modesto 1995.)

vertebrae, and these vertebrae are relatively short. The available skeletal evidence indicates that diadectids were herbivores, and this feeding strategy evolved within the Diadectomorpha. Berman *et al.* (1998) noted that the teeth of juvenile diadectids were much less molariform and had much more limited occlusion than those of the adults, possibly indicating an ontogenetic shift from omnivory to herbivory (or from low-fiber to high-fiber herbivory).

Synapsida: Edaphosauridae

Another early herbivore is the Permo-Carboniferous synapsid *Edaphosaurus* (Figures 2.2B, 2.3B) from North America (Romer and Price 1940; Reisz 1986). It is best known for its spectacular dorsal 'sail' supported by the greatly elongated dorsal neural spines, which are studded with prominent lateral tubercles. Unlike the condition in related, presumably faunivorous, forms its skull is small relative to overall body size. The crowns of the anterior marginal teeth of *Edaphosaurus* have distinct cutting edges, which are set obliquely relative to the long axis of the tooth row, and are suitable for cutting off pieces of plants (Modesto 1995). The cutting edges

Figure 2.3. Skeletons and body silhouettes (in black) of (A) *Cotylorhynchus* and (B) *Edaphosaurus*. The distal portion of the tail was omitted in both restorations for layout purposes. (Original drawings by D. M. Scott.)

of unworn teeth bear fine, oblique serrations, which subsequently became obliterated by wear. The palate (Figure 2.2B) and the lingual surfaces of the mandibular rami both support massive tooth plates, each bearing several rows of numerous teeth. Wear on these teeth indicates that food was crushed between the opposing upper and lower tooth plates by occlusion and by fore-and-aft motion of the mandible (Modesto 1995; Hotton *et al.* 1997). The latter is indicated by the considerable size disparity between the articular surfaces of the quadrate and the articular bone; the latter is about 50% longer anteroposteriorly than the former (Romer and Price 1940). The large, barrel-shaped trunk of *Edaphosaurus* is consistent with the presence of a voluminous digestive tract suitable for fiber fermentation.

Modesto (1994) developed a phylogenetically constrained scenario for the evolution of feeding habits in Edaphosauridae. The most basal taxon known, the Late Pennsylvanian *Ianthasaurus* (Figure 2.2A), has a dentition comprising slightly recurved, labiolingually compressed teeth that is consistent with faunivorous, probably insectivorous habits (Modesto and

Reisz 1990). The more derived Early Permian *Glaucosaurus* retains labiolingually compressed, sharp teeth, but its snout is much shorter, its dentition is isodont, and an antorbital buttress is developed. Modesto (1994) argued that these features indicate a more omnivorous diet, perhaps including harder food items. Unfortunately, *Glaucosaurus* is currently known only from a single incomplete skull, and thus comparisons have to be restricted to *Ianthasaurus* and other basal synapsids.

Comparisons between *Edaphosaurus* and *Ianthasaurus,* as well as with other eupelycosaurs, indicate that the following cranial features may also be related to the adoption of herbivory in the former: reduced ratio of skull length to trunk length, foreshortened snout, greatly enlarged adductor chamber as a result of changes in the shape of the squamosal and a shift of the jaw joint to a more ventral position, and massive, deep lower jaw. In the postcranial skeleton, the cervical vertebrae are markedly smaller than the trunk vertebrae, possibly in relation to the relatively small head of these forms, and the trunk region is shorter (with three fewer vertebrae) but has a wider rib cage with more strongly curved ribs than in related faunivorous forms. The fossil record provides compelling evidence that herbivory evolved within the Edaphosauridae.

Synapsida: Caseidae

The Caseidae, from the Permian of North America and Europe (Olson 1968; Reisz 1986), are generally interpreted as herbivores. Traditionally considered close relatives of Edaphosauridae (Romer and Price 1940), more recent phylogenetic analyses (e.g., Reisz 1986) have reinterpreted them as part of the basal clade of Synapsida, Caseasauria. Unlike the condition in faunivorous basal synapsids and outgroup taxa such as *Limnoscelis,* all caseids have a very small head relative to overall body size (Figure 2.3A). The unusually broad and barrel-shaped trunk indicates the development of an enormous digestive tract. The anterior teeth are the largest ones, and may have been used to crop plant material (Figure 2.4B). Their crowns are anteroposteriorly compressed, spatulate, and frequently lack apical cusps. The more posterior teeth have labiolingually compressed crowns that become spatulate toward the apices and bear prominent apical cusps or denticles (Olson 1968). They closely resemble the teeth of extant plant-eating iguanid lizards. The number of apical cusps varies between individual taxa, ranging from three to eight (Reisz 1986). Although the upper and lower teeth did not meet in occlusion, they could puncture and shred plant matter. In marked contrast to the condition in iguanid lizards, the palatal dentition is very well developed in Caseidae,

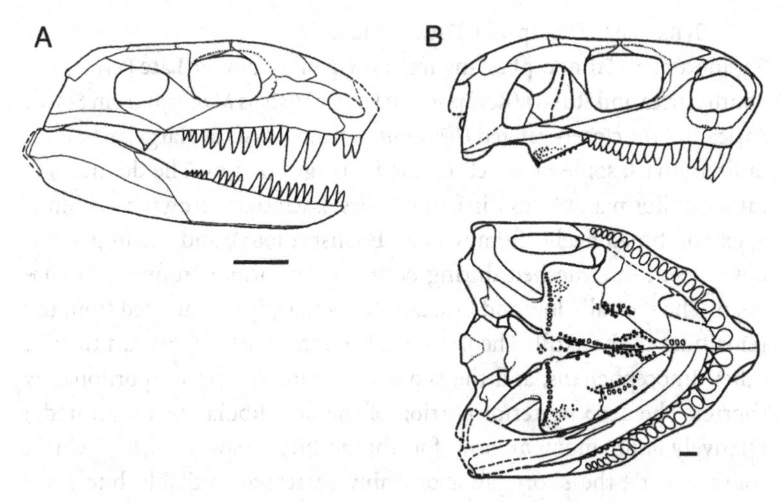

Figure 2.4. Skulls of (A) *Eothyris* and (B) *Cotylorhynchus* in lateral and palatal views. The palate of *Eothyris* is only partially known, preventing a detailed reconstruction. Scale bars each equal 1 cm. (Original drawings by D. M. Scott.)

ranging from the primitive pattern seen in *Cotylorhynchus* (with transverse, diagonal, and longitudinal rows of palatal teeth) to the highly derived condition seen in *Casea* (with a massive pavement of teeth covering most of the palate). Furthermore, the palatal dentition extends onto the ventral surface of the parasphenoid and cultriform process in all caseids. The large, well-ossified hyoid apparatus indicates the development of a substantial tongue for manipulating food within the oral cavity (Olson 1968). The tongue could have positioned and moved food against the palatal dentition. The jaw joint is situated below the level of the lower tooth row. The forelimbs are robust and unusually large, and the hands terminate in large claws suitable for digging (Stovall, Price and Romer 1966). *Cotylorhynchus romeri*, from the Early Permian of Oklahoma, reached a length of about 3.5 m and an estimated live weight of up to 331 kg (Stovall *et al.* 1966), but *C. hancocki*, from slightly younger Permian strata in Texas, was up to 25% larger in comparable linear dimensions (Reisz 1986).

The Eothyrididae, the sister-taxon of Caseidae among Caseasauria (Reisz 1986), are characterized by a distinctly heterodont dentition composed of simple conical marginal teeth (Figure 2.4A), which indicate faunivorous habits. Outgroup comparisons are not possible for postcranial features because the two known eothyridid genera (*Eothyris* and *Oedaleops*) are represented only by cranial remains. Herbivory evolved within Caseasauria, and even the most basal caseids already fed on plants.

Synapsida: Therapsida: Dinocephalia

Tapinocephalid dinocephalians are known only from the Late Permian of South Africa and Russia (Kemp 1982). Forms such as *Moschops* from South Africa and the closely related *Ulemosaurus* from Russia are large, massively built animals, some of which reached a length of 3 m. The dentition is rather uniform and lacks distinct canines. Each tooth crown has a blunt apex and broad heel (Efremov 1940; Boonstra 1962), and the upper and lower teeth intermeshed during occlusion to form a cropping mechanism. The typically low and broad snout is sharply demarcated from the remainder of the skull. The jaw joint is located further forward than in other dinocephalians, and the snout and mandible are proportionately shorter. The deep posterior portion of the mandibular ramus created a relatively long moment arm for the adductor jaw muscles, which, coupled with the short jaws, probably increased available bite force (Kemp 1982). The rib cage is unusually large, and the transverse processes of the dorsal vertebrae are long. The trunk probably housed a voluminous digestive tract (Gregory 1926).

The Tapinocephalidae represent a derived clade of presumed herbivores within Dinocephalia, which are typically and plesiomorphically carnivorous.

Synapsida: Therapsida: Anomodontia

Anomodont therapsids (Figure 2.5), especially the Dicynodontia, were the most successful group of Permian continental tetrapods in terms of both taxonomic diversity and overall abundance (King 1990, 1996). Dicynodonts ranged from the Late Permian to the Late Triassic and had a worldwide distribution. They are most common, both in terms of number of taxa and especially number of individuals, in the Late Permian continental strata of the Beaufort Group in South Africa (Kitching 1977; King 1990, 1996). The largest forms have skull lengths exceeding 50 cm. Some Middle and Late Triassic dicynodonts may have exceeded 3 m in length and weighed one metric ton or more. The entire skeleton is robustly built. The skull (Figure 2.5C) is highly derived compared with that in other therapsids (Crompton and Hotton 1967; Cluver 1971; King 1990, 1996). Most derived dicynodonts have enlarged upper caniniform teeth but lack postcanine teeth. The anterior ends of the snout and mandible were probably covered by an extensive keratinous beak (rhamphotheca) in life, and the lower beak bit inside the upper when the mouth was closed. Judging from the underlying bony ridges, it seems that the

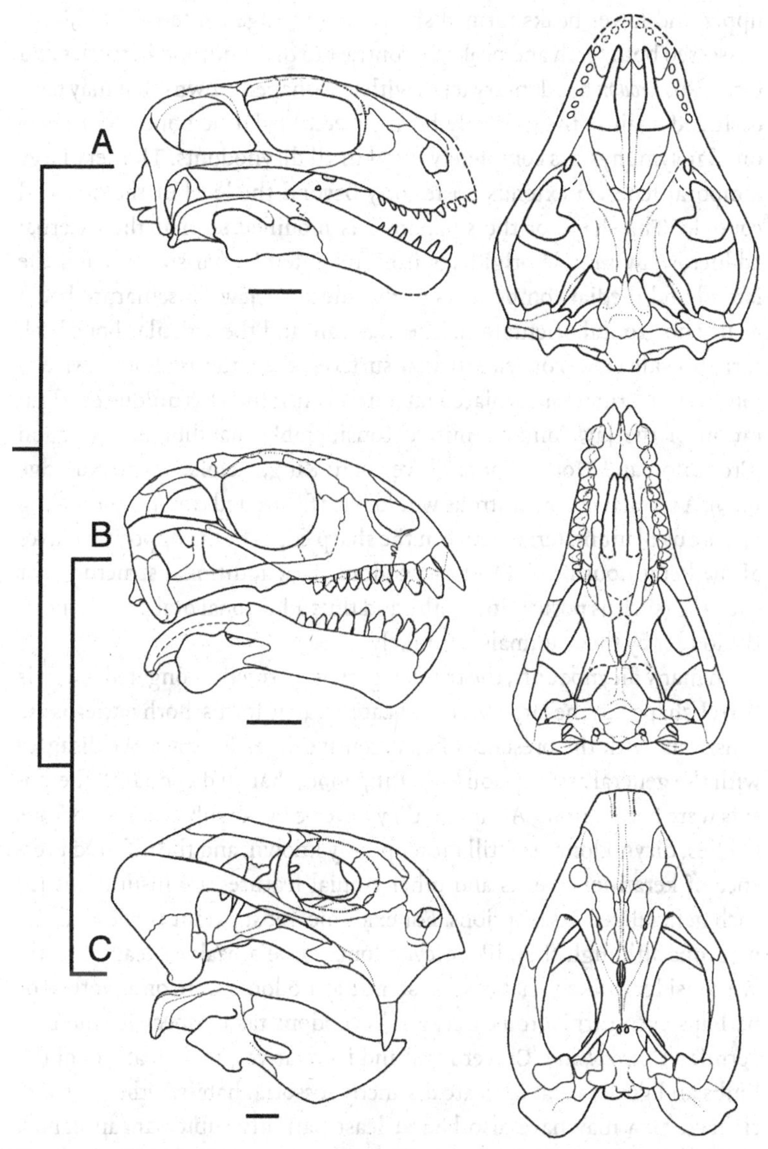

Figure 2.5. Skulls of three representatives of Anomodontia ((A) *Patranomodon*, (B) *Suminia*, and (C) *Dicynodon*) in lateral (with attached mandible) and palatal views, arranged on a cladogram depicting a hypothesis of the interrelationships of these taxa. Scale bars each equal 1 cm. (*Patranomodon* redrawn from Rubidge and Hopson 1996, courtesy and copyright of the Linnean Society of London; *Suminia* courtesy and copyright of N. Rybczynski; *Dicynodon* redrawn from Cluver and King 1983, courtesy and copyright of the South African Museum.)

upper and lower beaks formed sharp cutting edges. A few dicynodonts possessed both teeth and beaks, in contrast to the condition in turtles and birds. *Pristerodon* has dentary teeth with leaf-shaped crowns that may have occluded against the (possibly horn-covered) palatine bones. The mandibular symphysis is completely fused in all dicynodonts. The very large temporal fenestra extends posteriorly beyond the level of the occipital condyle. The shape of the squamosal is modified so that the external adductor jaw muscle originates from its anterolateral surface, and the lateral and medial components of the adductor jaw musculature had a nearly horizontal orientation. The quadrate and the articular bone both form (in side view) convex articular surfaces; the latter is almost twice as long as the former and is placed far anteroventrally. This unique configuration of the jaw joint permitted considerable mandibular retraction (Crompton and Hotton 1967; Cluver 1971; King, Oelofsen, and Rubidge 1989). A retractive power stroke would have facilitated cutting and slicing of fibrous plant material between the sharp edges of the opposing halves of the keratinous beak. Most dicynodonts lack additional structures for chewing (such as postcanine teeth), and thus additional oral processing of the fodder by these animals is unlikely.

In many dicynodonts, the trunk region is distinctly elongated, but it is barrel-shaped in the large Mid- and Late Triassic forms; both patterns are consistent with the presence of an extensive digestive tract. We disagree with the general assumption (e.g., King 1990) that all dicynodont therapsids were herbivorous. Although they have been the subject of numerous studies, dicynodonts are still rather poorly known, and the inferred presence of keratinous beaks and other cranial features are insufficient for such generalized assumptions. For example, most extant turtles are carnivorous although they, like dicynodonts, have a beak instead of teeth. We consider it likely that at least some dicynodonts were omnivorous or perhaps even carnivorous. Certain dicynodont taxa, especially the Late Permian *Cistecephalus* (Cluver 1978) and its relatives, show features of the limbs and girdles that indicate distinctly fossorial habits. Other less specialized taxa may have also had at least partially subterranean habits. Skeletons of the Late Permian *Diictodon* have been found in the terminal chambers of complex helical burrows (Smith 1987).

More basal anomodonts differ from dicynodonts especially in the presence of well-developed marginal teeth. *Patranomodon* (Figure 2.5A) from the lowest of the Permian-age strata of the Beaufort Group of South Africa is considered the most basal anomodont (Rubidge and Hopson 1996). Its dentition is still poorly known, but it apparently lacks any specializations for herbivory. In the absence of such dental features, it is difficult to evaluate the feeding habits of this small anomodont, but some of its cranial features, such as the short snout, are generally found in herbivorous tetrapods. *Patranomodon* lacks the typically dicynodont jaw articulation and has a screw-shaped articular surface between quadrate and articular, which precludes fore-and-aft motion of the mandible. In contrast, *Suminia* (Figure 2.5B) from the Late Permian of Russia has teeth with leaf-shaped, cuspidate crowns (Rybczynski 1996). Striations on the well-developed wear facets demonstrate that jaw motion comprised mandibular elevation and retraction. Many cranial features typical of later dicynodonts are already present in this small anomodont. For example, the squamosal flares out posterolaterally, providing an area for the origin of the external adductor jaw muscle (Crompton and Hotton 1967). As the sister-taxon of Dicynodontia, *Suminia* provides evidence that many cranial characters found in later dicynodonts are not necessarily related to the origin of herbivory in that group. Most significant among these features is the presence of a keratinous beak, the much enlarged, dorsolaterally oriented temporal fenestra, and the partial development of a secondary bony palate. In dicynodonts, these features may represent further refinements for feeding on plants. The fossil record provides compelling evidence that herbivory evolved within the clade Anomodontia.

Reptilia: Parareptilia: Pareiasauria

The Pareiasauria included some of the largest Paleozoic herbivorous tetrapods. These bulky reptiles range in length from about 1 m to 3.5 m and have large, massive skulls (Bystrov 1957; Lee 1997). They appear to have been restricted to the Late Permian and had an almost worldwide distribution. The snout is transversely broad, and the postorbital region is much expanded transversely (Figure 2.6). Each jaw bears a single marginal row of closely spaced, non-occluding teeth with crowns that superficially resemble those of caseid synapsids in being tall, strongly flattened labiolingually, and blade-like towards the apex of the crown. The apices of the tooth crowns each bear six or more cusps or denticles and are lined up along the entire length of the jaw to form a continuous shearing edge. The number of cusps increased significantly and the tooth crowns became fan-shaped in more derived pareiasaurs (Lee 1997). The large, broad trunk of pareiasaurs is consistent with the presence of a voluminous digestive tract. Although the phylogenetic relationships of Pareiasauria are still

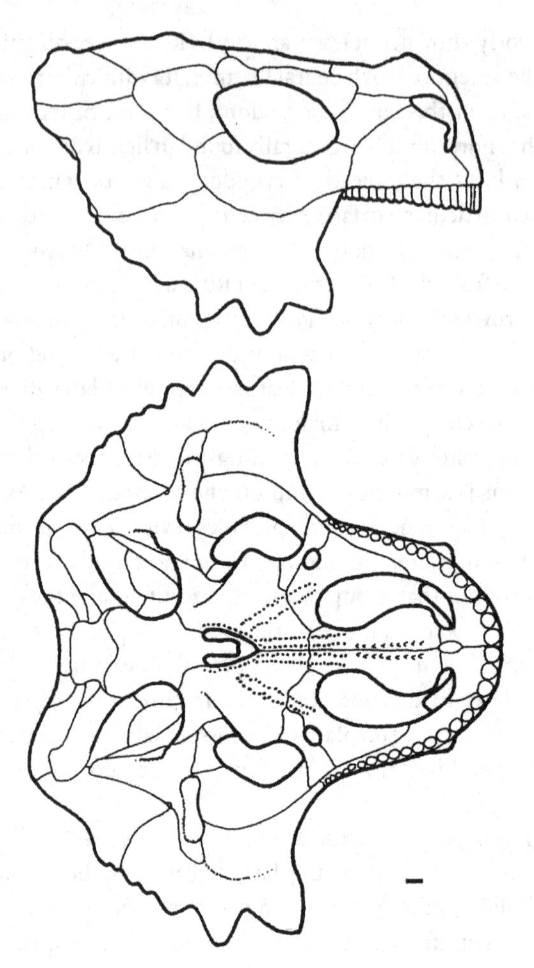

Figure 2.6. Skull of a pareiasaur, *Scutosaurus*, in lateral and palatal views. Scale bar equals 1 cm. (Redrawn from Bystrov 1957.)

somewhat controversial, the Procolophonoidea are the most likely sister-group of this taxon among Parareptilia (Laurin and Reisz 1995), with small faunivorous forms such as *Owenetta* (see Figure 2.10) presumably representing the plesiomorphic condition for that group. Using this hypothesis of relationships for outgroup comparisons, the following cranial features of pareiasaurs may be related to herbivory: short tooth rows and snout, posterodorsally expanded skull table, and reduced palatal dentition. In the postcranial skeleton, the vertebral column is reduced to 19 or 20 presacral vertebrae (Lee 1997) and stiffened, with each

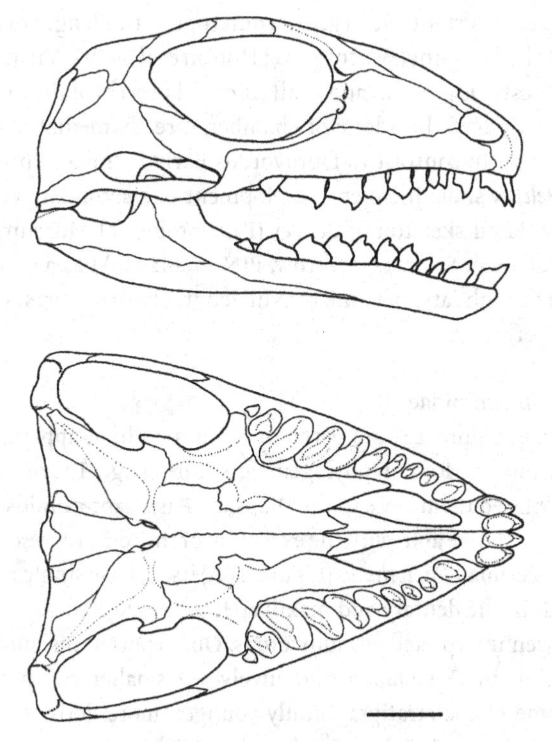

Figure 2.7. Skull of a bolosaurid, *Belebey*, in lateral and palatal views. Scale bar equals 1 cm. (Modified from Ivakhnenko and Tverdokhlebova 1987 and specimens.)

dorsal vertebra being short but tall and massively built. The ribs are large and strongly curved.

Reptilia: Bolosauridae

The Bolosauridae form a clade of still poorly known basal reptiles from the Early Permian of North America (*Bolosaurus*; Watson 1954, Hotton *et al.* 1997) and the Late Permian of Russia and China (*Belebey*; Ivakhnenko and Tverdokhlebova 1987, Li and Cheng 1995). The premaxillary and anterior dentary teeth are somewhat procumbent (Figure 2.7). The crowns of the maxillary teeth of *Bolosaurus* are bulbous labially, with a single robust, lingually curved cusp. A prominent cingulum extends lingual to the base of the cusp and delimits a shallow basin between it and the base of the labial cusp. Similarly, the dentary teeth bear a prominent lingual cusp and a labial cingulum. The dentary and maxillary teeth are thickly enamelled. Striations on the distinct wear facets indicate fore-and-aft motion of the mandible, which is consistent with a greater anteroposterior length of the articular facet of the jaw joint (Watson 1954; Hotton *et al.* 1997). A narrow infratemporal fenestra is present, and a tall coronoid process of the lower jaw extends dorsally into the adductor chamber, directly medial to the fenestra (Figure 2.7). In contrast to faunivorous Early Permian reptiles, *Bolosaurus* and *Belebey* show incipient development of a secondary bony palate. One articulated skeleton of *Belebey* (Paleontological Museum of the Russian Academy of Sciences, Moscow, PIN 104B/2020) has a barrel-shaped trunk and a high ratio of trunk to skull length, both features commonly found in herbivores.

Reptilia: Captorhinidae

The Captorhinidae (Figure 2.8) are Permian reptiles with an apparently worldwide distribution. Recent phylogenetic studies (e.g., Laurin and Reisz 1995) have placed them close to the Diapsida. Basal captorhinids are characterized by small size and, with the exception of the some representatives of the Early Permian *Captorhinus* (Figure 2.8A) itself, have single rows of marginal teeth in the dentary and maxilla (Heaton 1979). These forms have generally been interpreted as faunivorous. One reported instance of apparent predation in *Captorhinus aguti* involves a smaller conspecific (Eaton 1964). Some of the stratigraphically younger, more derived, and larger captorhinids such as *Labidosaurikos* (Figure 2.8C) have certain craniodental features that are consistent with feeding on plants (Dodick and Modesto 1995). *Labidosaurikos* and the Late Permian *Moradisaurus* differ from other captorhinid reptiles in being much larger (with the skull length of *Moradisaurus* exceeding 40 cm), and having broad dentaries and maxillae that bear multiple (6 to 11) rows of teeth. The small, rather isodont teeth are set in parallel longitudinal rows. The tooth crowns show distinct labial and lingual wear facets. Both the pattern of tooth wear (Dodick and Modesto 1995; Hotton *et al.* 1997) and the structure of the jaw joint (Ricqlès and Taquet 1982) indicate fore-and-aft motion of the mandible, and the lower rows of teeth fit neatly between the upper rows for effective crushing and shredding of plant material. The postcranial skeleton of *Labidosaurikos* is still unknown, and that of *Moradisaurus* has yet to be described. Therefore, we restrict our comparisons to cranial features. The sister-taxon of the presumably herbivorous Moradisaurinae is the single-tooth-rowed *Labidosaurus* (Figure 2.8B). It is difficult to determine the feeding habits of the latter, but there exists no clear evidence to suggest that it was a herbivore. Comparisons with this form and other more basal captorhinid taxa indicate that the Moradisaurinae have additional skeletal features that may be

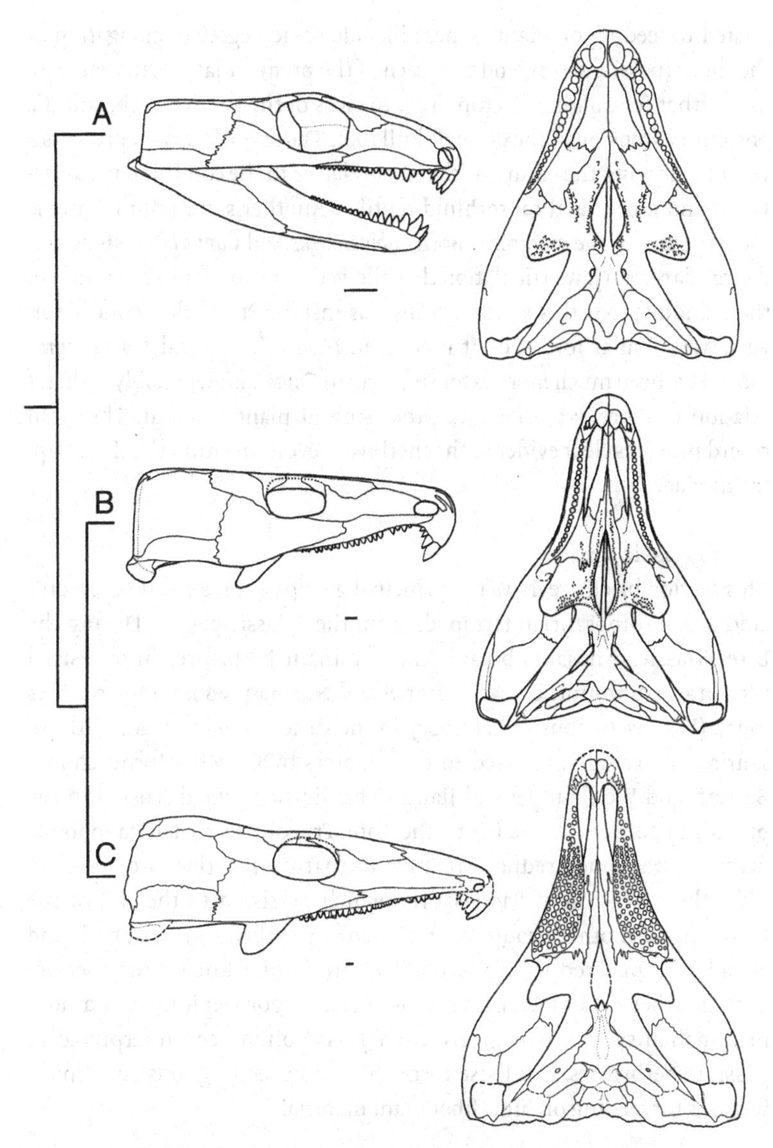

Figure 2.8. Skulls of three representatives of Captorhinidae (A, *Captorhinus*; B, *Labidosaurus*; and C, *Labidosaurikos*) in lateral and palatal views, arranged on a cladogram depicting a hypothesis of the interrelationships of these taxa. *Captorhinus* includes both forms with single (*C. laticeps*, depicted here) and multiple marginal tooth rows (*C. aguti*). The hypothesis of relationships depicted here postulates that multiple tooth rows evolved twice within this clade (Dodick and Modesto 1995). Scale bars each equal 1 cm. (*Captorhinus* modified from Heaton 1979 and specimens; *Labidosaurus* original drawing by D. M. Scott; *Labidosaurikos* modified from Dodick and Modesto 1995 and specimen.)

related to feeding on plants. These include some degree of elongation of the snout (possibly to extend the reach of the premaxillary teeth that were used either for digging or cropping), the loss of the palatal teeth, and the posterior extension of the dermal skull roof. The latter feature represents a way of enlarging the adductor chamber relative to the condition in *Labidosaurus* and more basal captorhinid reptiles. Another significant feature is the structure of the jaw joint. Both *Labidosaurus* and *Captorhinus* show evidence of an open jaw articulation that allowed some fore-and-aft motion of the mandible, but this feature by itself is insufficient evidence for herbivory. Mandibular fore-and-aft motion in *Labidosaurikos* and *Moradisaurus* must have been much more extensive than in *Captorhinus*, possibly in direct relation to more extensive oral processing of plant material. The fossil record provides clear evidence that herbivory evolved within the clade Captorhinidae.

Triassic taxa

This section briefly reviews the principal groups of presumed herbivores among non-dinosaurian tetrapods from the Triassic period. During the Late Triassic, dinosaurs became the dominant herbivores in terrestrial vertebrate communities (Weishampel and Norman 1989; Wing and Sues 1992). (For discussions of herbivory in the various major clades of Dinosauria, the reader is referred to the chapters by Barrett, Upchurch and Barrett, and Weishampel and Jianu.) The dicynodonts, discussed in the preceding section, ranged into the Late Triassic, but their taxonomic diversity was much reduced relative to that during the Late Permian. With the exception of Pareiasauria, which persisted to the end of the Permian, most other groups of late Paleozoic herbivorous tetrapods had already disappeared from the fossil record earlier during that period. Certain other taxa of Triassic reptiles (e.g., Stagonolepididae) and non-mammalian synapsids (e.g., Bauriidae) have often been interpreted as possible herbivores, but these forms have no unambiguous anatomical features for feeding on high-fiber plant material.

Synapsida: Cynodontia: Gomphodontia

The Gomphodontia formed a diverse clade of mostly Triassic cynodonts (Figure 2.9B). They are readily characterized by the possession of transversely expanded, molariform postcanine teeth that meet in complex tooth-to-tooth occlusion (Crompton 1972). Some derived forms, such as the traversodont *Exaeretodon* from the Late Triassic of Argentina and India, may have reached a length of up to 2 m. The skull is large and

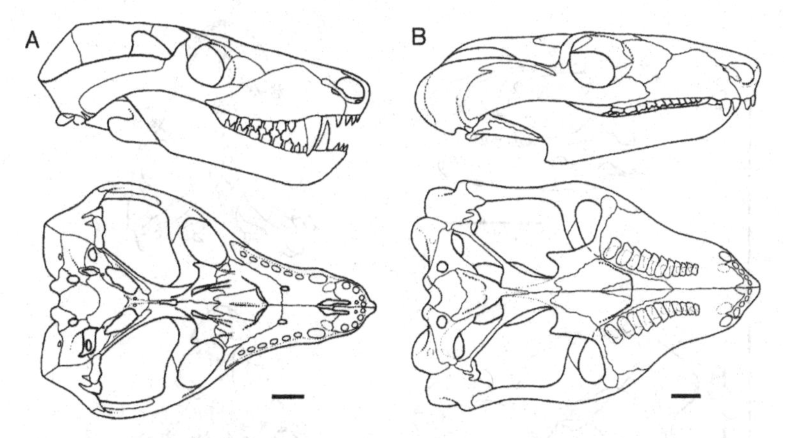

Figure 2.9. Skulls of two representatives of Cynodontia (A, *Thrinaxodon* and B, *Massetognathus*) in lateral and palatal views. Scale bars each equal 1 cm. (*Thrinaxodon* modified from Parrington 1946 and specimens; *Massetognathus* modified from Romer 1967 and specimens.)

massive, with deep zygomatic arches and often pronounced ectocranial crests. The rostral portions of the dentaries are fused to form an extensive mandibular symphysis. In derived traversodont taxa, the tightly packed crowns of the molariform upper and lower postcanine teeth each bear a transverse crest. Each molariform forms multiple shearing surfaces. Occlusion was bilateral and involved a power stroke during which the shearing surfaces of the lower postcanine teeth were dragged posterodorsally across the matching surfaces on the upper teeth (Crompton 1972). The postcanine teeth are distinctly inset from the sides of the face, possibly indicating the development of a lateral oral vestibule bounded by muscular cheeks (Hopson 1984). The temporal fossae are much enlarged and delimited by prominent ectocranial crests in derived gomphodont cynodonts, and the dentary forms a tall, often recurved coronoid process, indicating powerful development of the temporalis musculature.

The dentitions of the Early Triassic basal gomphodonts *Diademodon* and *Trirachodon* are distinctly heterodont, comprising conical anterior teeth, transversely expanded molariforms, and sectorial posterior teeth (Hopson 1971; Crompton 1972; Grine 1977). The multicuspid sectorial teeth are very similar to the postcanine teeth of faunivorous eucynodonts such as *Cynognathus* and *Thrinaxodon* (Figure 2.9A). The pattern of dental microwear indicates that *Diademodon* retained orthal jaw motion (Grine 1977). Based on the heterodont nature of the postcanine dentition, basal gomphodont cynodonts probably were omnivorous, and feeding on

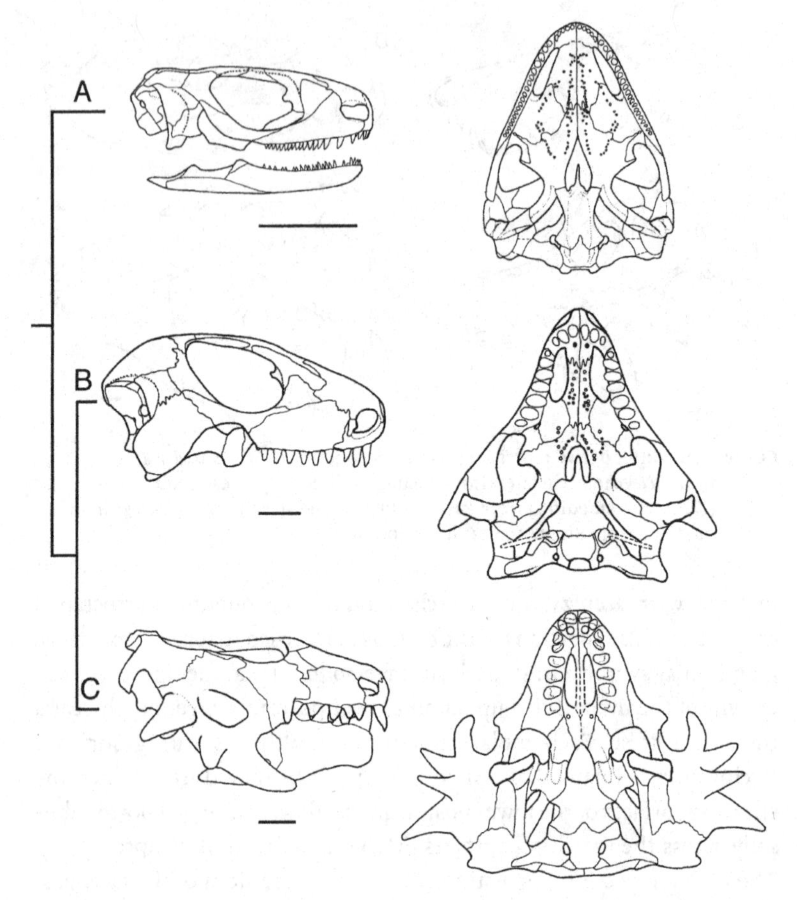

Figure 2.10. Skulls of three representatives of Procolophonoidea (A, *Owenetta*; B, *Procolophon*; and C, *Hypsognathus*) in lateral and palatal views, arranged on a cladogram depicting a hypothesis of the interrelationships of these taxa. Scale bars each equal 1 cm. (*Owenetta* original drawing by D. M. Scott; *Procolophon* modified from Carroll and Lindsay 1985 and specimens; *Hypsognathus* modified from Sues *et al.* 2000).

high-fiber plant material evolved within the clade Gomphodontia. This heterodont type of dentition was even retained in juvenile specimens of some traversodont taxa (H.-D. Sues, unpublished data), and may reflect an ontogenetic shift from more omnivorous habits in juveniles to a predominantly herbivorous diet in adults.

Reptilia: Parareptilia: Procolophonidae

The Procolophonidae are small to medium-sized, superficially lizard-like parareptiles (Figure 2.10). They had a worldwide geographic distribution and spanned the entire Triassic period. The skull is particularly characterized by a distinct posterior embayment of the orbit, which probably accommodated substantial adductor jaw muscles. This embayment is most pronounced in the highly derived Late Triassic *Hypsognathus* (Figure 2.10C) and *Leptopleuron* (Leptopleuroninae), in which the opening is expanded posteriorly as well as laterally. The lower jaw in these forms also has a tall, occasionally recurved coronoid process for the insertion of the adductor jaw muscles, and the jugal is much deeper. The jaw joint in *Hypsognathus* and other derived leptopleuronines is situated well below the level of the lower tooth row. The dentition of procolophonids is differentiated into incisiform anterior teeth and transversely widened, bicuspid or multicuspid posterior teeth. On unworn posterior teeth, pointed labial and lingual cusps are linked by a sharp transverse crest, which became progressively obliterated by wear to form a more or less flat apical wear facet (Gow 1978; Carroll and Lindsay 1985; Sues and Baird, 1998). In most taxa, the maxillary and posterior dentary teeth interdigitated in a cog-like fashion. A noteworthy postcranial feature in some procolophonid taxa, such as *Hypsognathus,* is the transversely broad rib cage.

The Owenettidae (*Barasaurus* and *Owenetta*; Figure 2.10A) are generally considered the sister-group of Procolophonidae (Gow 1978; Laurin and Reisz 1995). They have conical, slightly recurved marginal teeth of fairly uniform size, with the lower teeth biting inside the upper ones. This type of dentition is consistent with probably insectivorous habits (Gow 1978). In addition, the vertebrae are not as robustly constructed as those of similar-sized procolophonids, and the trunk is much more slender than in *Hypsognathus* and *Procolophon*. Gow (1978) argued that juveniles of *Procolophon* (Figure 2.10B) and smaller procolophonid taxa were probably insectivorous, and only the adults of the larger forms with transversely expanded teeth were more or less exclusively herbivorous. The derived leptopleuronines (e.g., *Hypsognathus*; Figure 2.10C) show a number of features consistent with feeding on high-fiber plant material, such as teeth suited for shearing, a ventrally off-set jaw joint, and powerful development of the adductor jaw muscles. Herbivory thus evolved within the Procolophonidae.

Reptilia: Diapsida: Archosauromorpha: Rhynchosauria

The Rhynchosauria (Figure 2.11) are a derived group of Triassic archosauromorph reptiles with a nearly worldwide distribution. They were particularly abundant in the Middle and Late Triassic of Gondwana. Rhynchosaurs are characterized by the distinctive structure of the skull.

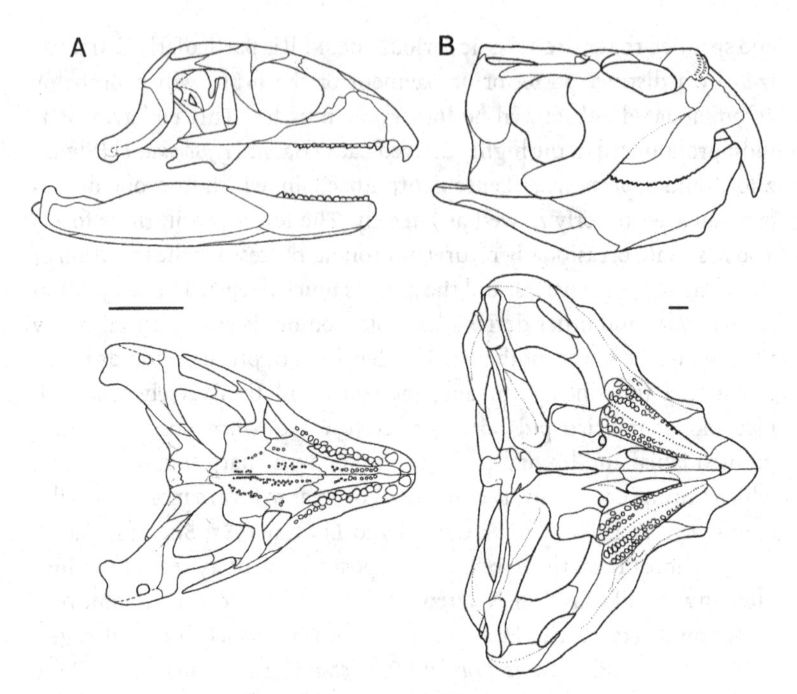

Figure 2.11. Skulls of two representatives of Rhynchosauria (A, *Mesosuchus* and B, *Hyperodapedon*) in lateral and palatal views. Scale bars each equal 1 cm. (*Mesosuchus* courtesy of D. W. Dilkes; *Hyperodapedon* redrawn from Benton 1983, courtesy and copyright of the Royal Society of London.)

In the most derived taxa such as *Hyperodapedon*, the transverse width of the temporal region exceeds the anteroposterior length of the skull (Benton 1983; Figure 2.11B). The temporal fossae are greatly enlarged, providing space for powerful adductor jaw muscles. The premaxilla is beak-like and devoid of teeth, and the rostral tip of the dentary also lacks teeth. The maxillae bear large tooth plates with multiple longitudinal rows of short conical teeth on either side of a median groove. The teeth are deeply rooted and firmly attached with bone (ankylothecodont implantation; Benton 1984). The ridge formed by the dentary teeth fits into the median groove on the maxillary tooth plate. Jaw motion was orthal and primarily involved shearing, with the food being cut as the lower jaws were brought diagonally into occlusion (Benton 1983, 1984). Rhynchosaurs are generally considered herbivorous; earlier suggestions of molluscivorous habits are inconsistent with the structure of the jaw apparatus and the pattern of tooth wear (Benton 1983). The more derived Rhynchosauridae are characterized by the possession of a broad, barrel-shaped trunk, which indicates the presence of a voluminous digestive tract (Benton 1983). Some taxa such as *Hyperodapedon* have unusually deep, mediolaterally flattened unguals, which may have been used for scratch-digging to recover underground vegetal structures such as rhizomes and tubers (Benton 1983).

The basal rhynchosaur *Howesia* from the Early or early Middle Triassic of South Africa has multiple rows of ankylothecodont teeth in the maxilla and dentary, but the upper and lower teeth met along a broad occlusal surface, rather than in the shearing fashion characteristic of more derived forms (Benton 1984; Dilkes 1995). The even more primitive *Mesosuchus* (Figure 2.11A) has a zigzag arrangement of marginal teeth in the dentary and maxilla, rather than a single continuous marginal tooth row in each jaw (Dilkes 1998).

The sister-taxa of Rhynchosauria are faunivorous with the exception of *Trilophosaurus*, which exhibits rather different craniodental specializations (see below). The distinctive jaw configuration for shearing and thus presumably herbivory evolved within Rhynchosauria because *Howesia* and *Mesosuchus* still lack this configuration.

Reptilia: Diapsida: Archosauromorpha: Trilophosauridae

Trilophosaurus from the Late Triassic of Texas (Gregory 1945; DeMar and Bolt 1981) has posterior teeth with relatively high and transversely much expanded crowns. The premaxilla and anterior end of the dentary are devoid of teeth and may have been covered by a keratinous beak in life. With the exception of a few anterior single-cusped teeth, the dentary and maxilla bear teeth with transversely expanded crowns, each with three subequal cusps linked by a transverse crest on unworn teeth. The transversely expanded teeth typically show extensive wear. Jaw motion was orthal and crushing, with the upper and lower teeth interdigitating in a cog-like fashion (Gregory 1945). The adductor jaw musculature was substantial, as indicated by the deep temporal region and extensive temporal fossa, which is laterally enclosed by a deep, unfenestrated temporal bar.

Trilophosaurus is generally referred to Archosauromorpha (see Benton 1983), but its precise phylogenetic position within that clade is still unresolved. Most other archosauromorph reptiles, and Diapsida plesiomorphically, are faunivorous.

Summary

The preceding overviews of presumed plant-eating tetrapods from the late Paleozoic and Triassic (with the exception of dinosaurs) establish two basic dental patterns for herbivory. The first, found in most of the taxa discussed above, is characterized by dentitions where occlusion between the upper and lower teeth breaks down food material. This is the most common pattern, appearing first in the fossil record, and shows the greatest variation in the development of structures used for oral processing. Rhynchosaurid reptiles show a particularly intricate mechanism for puncture crushing and shearing. Gomphodont cynodonts developed molariform teeth with complex cusp patterns and precise tooth-to-tooth occlusion.

The second pattern, which first developed in caseid and basal anomodont synapsids as well as in pareiasaurs, is characterized by the presence of leaf-shaped, cuspidate marginal teeth suitable for puncturing and tearing apart fibrous plant material, in a manner similar to that seen in extant iguanid lizards. It probably implies less extensive oral processing of food than the first pattern of dentition. (The beaks of dicynodont therapsids may have served a similar function.) In all these forms, the anterior marginal teeth presumably served to seize and cut off parts of plants for oral processing and ingestion. Among Triassic tetrapods, leaf-shaped, cuspidate tooth crowns are known only in prosauropod and basal ornithischian dinosaurs.

Compared with their faunivorous relatives, both types of Paleozoic and Triassic herbivorous tetrapods show various cranial features suitable for enhanced jaw action, most importantly a relative increase in the size and thus power of the adductor jaw muscles. This change is reflected by either enlargement of the adductor chamber of the skull in forms with closed temporal regions (Captorhinidae, Pareiasauria) or by an increase in the size of the temporal opening in forms with a fenestrated temporal region (such as Synapsida, which have a single temporal opening on either side of the skull; Figure 2.3). In the Dicynodontia and their relatives (Figure 2.4), the external adductor jaw musculature further increased in size through expansion onto the ventrolateral surface of the squamosal. Although they differ considerably in their craniodental structure, dicynodonts and gomphodont cynodonts both employed a palinal power stroke suitable for shearing.

Most of the presumed herbivores discussed above are characterized by medium to large body size and capacious trunks suitable for accommodating extensive digestive tracts for endosymbiotic cellulysis. This is consistent with the observation that large body size facilitates more efficient processing of high-fiber plant fodder by cellulysis (Farlow 1987; Hotton *et al.* 1997). The mass-specific metabolic rate of an animal decreases with increasing body size, whereas the ratio of gut capacity to body size remains more or less constant (Farlow 1987). A larger herbivore should have a lower turnover rate of its gut contents than a smaller one. Indeed, the rate of passage of food through the gut generally decreases with larger body size (Parra 1978), so that the ingested fodder can be exposed to microbial fermentation for longer periods of time and the yield from fermentation is higher because more cellulose is processed per unit of time.

Two basic types of body form can be distinguished among presumed early herbivores. In most Paleozoic and early Mesozoic herbivorous tetrapods, the rib cage is expanded laterally and even dorsally on either side of the vertebral column, resulting in a broad, barrel-shaped trunk. An apparent functional correlate of this change in many forms is the stiffening of the vertebral column by means of accessory intervertebral facets, greatly expanded zygapophyses, massive neural spines, and large facets for rib articulation. Another common feature in these forms is a reduction in the number of trunk vertebrae relative to their faunivorous sister-taxa. The most dramatic example of this condition is observed in the Pareiasauria. In the evolutionary history of this group, stiffening of the vertebral column may also have been linked to the development of dermal armor (Lee 1996), but this suggestion is not very convincing because the largest, bulkiest forms are the first to appear in the fossil record and show much less extensive development of armor than the smaller, stratigraphically younger, and more derived taxa. A second type of body form is represented by dicynodonts, many of which have a distinctly elongate trunk region. This represents an alternative condition to accommodate an extensive digestive tract.

Conclusions

The currently known fossil record indicates that adaptations for herbivory in terrestrial tetrapods appeared only some 60 million years after the invasion of land by vertebrates. However, once this novel feeding strategy had developed, it evolved independently in numerous lineages of terrestrial reptiles and non-mammalian synapsids (Figure 2.12). During the late Paleozoic, herbivory appears to have developed four times among non-mammalian synapsids, at least once among eureptiles, and twice among parareptiles. During the Triassic period, it evolved in at least one additional clade of synapsids and at least two groups of non-dinosaurian archosauromorph reptiles.

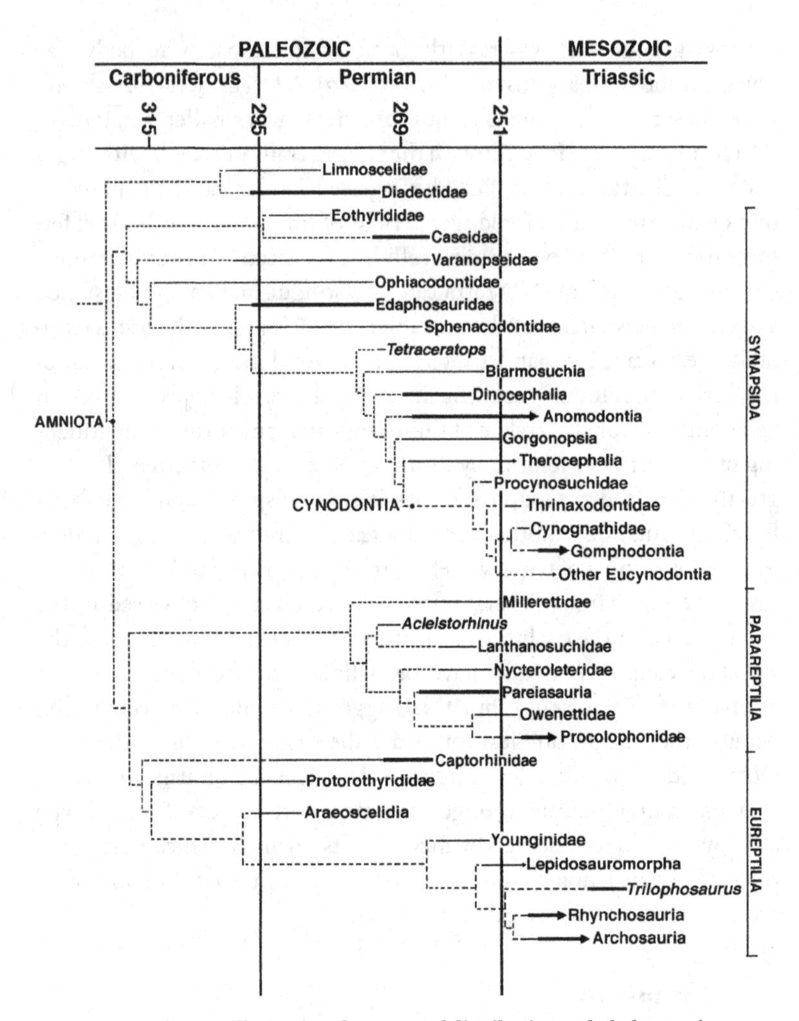

Figure 2.12. Diagram illustrating the temporal distribution and phylogenetic relationships of late Paleozoic and early Mesozoic amniotes (combined from numerous sources), illustrating the repeated, independent development of herbivory. Thick lines denote lineages of or including presumed herbivores; broken lines denote unrecorded ranges inferred from phylogenetic hypotheses.

Although some Late Carboniferous tetrapods already show craniodental features for feeding on high-fiber plant material, herbivorous forms did not appear in significant numbers until the Late Permian, some 40 million years after the first appearance of herbivores in the fossil record of terrestrial tetrapods. The reasons for this delay are not yet understood. The fossil record indicates that the structure of 'modern' terrestrial ecosystems apparently developed gradually, and their basic pattern of trophic interactions was only established during Late Permian times when herbivorous reptiles and synapsids first became an important component of the terrestrial food chain.

All Permo-Carboniferous and, with the exception of dinosaurs, all Triassic plant-eating tetrapods were quadrupedal animals that probably foraged within one meter above the ground. It is possible that some of the small forms may have also foraged in the trees, but none shows obvious specializations for a scansorial or arboreal mode of life.

The evolution of large body size in various lineages of herbivorous tetrapods appears to be linked to the acquisition of this mode of feeding. This trend is particularly well documented in captorhinid reptiles and in caseid and edaphosaurid synapsids. Late Permian tetrapod communities include, for the first time, several unrelated groups of large-bodied herbivores (Pareiasauria, Dinocephalia, Dicynodontia), some with a body mass in excess of 1000 kg. Large body size apparently evolved repeatedly among dicynodont therapsids (King 1990).

It would appear that the development of herbivory represented a major adaptive threshold in the evolution of terrestrial tetrapod communities. Once amniotes developed the ability to acquire and maintain colonies of cellulytic endosymbionts, this happened repeatedly in many lineages of amniotes, leading to the apparently gradual emergence of the trophic structure characteristic of modern terrestrial ecosystems.

Acknowledgments

Special thanks are due to D. M. Scott for preparing the figures. In many instances, her work on these illustrations included substantial reinterpretation of skull reconstructions from earlier publications through direct examination of specimens. D. W. Dilkes, S. P. Modesto, and N. Rybczynski kindly provided us with original drawings for use in this review. We are indebted to R. L. Carroll, N. Hotton III, and especially R. Beerbower for providing constructive comments on the manuscript. Both authors gratefully acknowledge continuing financial support for their research from operating grants from the Natural Sciences and Engineering Research Council of Canada (NSERC).

The sequence of authors' names is strictly alphabetical, and no seniority is implied.

References

Benton, M. J. (1983). The Triassic reptile *Hyperodapedon* from Elgin: functional morphology and relationships. *Phil. Trans. R. Soc. Lond. B* 302:605–717.

Benton, M. J. (1984). Tooth form, growth, and function in Triassic rhynchosaurs. *Palaeontology* 27:737–776.

Berman, D. S, Henrici, A. C., and Sumida, S. S. (1998). Taxonomic status of the Early Permian *Helodectes paridens* Cope (Diadectidae) with discussion of occlusion of diadectid marginal dentition. *Ann. Carnegie Mus.* 67:181–196.

Bjorndal, K. A. (1979). Cellulose digestion and volatile fatty acid production in the green turtle, *Chelonia mydas*. *Comp. Biochem. Physiol.* 63A:127–133.

Bjorndal, K. A., and Bolten, A. B. (1993). Digestive efficiencies in herbivorous and omnivorous freshwater turtles on plant diets: Do herbivores have a nutritional advantage? *Physiol. Zool.* 66:384–395.

Bjorndal, K. A., Bolten, A. B., and Moore, J. E. (1990). Digestive fermentation in herbivores: effect of food particle size. *Physiol. Zool.* 63:710–721.

Boonstra, L. D. (1962). The dentition of the titanosuchian dinocephalians. *Ann. S. Afr. Mus.* 46:57–112.

Bystrov, A. P. (1957). [Skull of pareiasaur.] *Trudy Paleont. Inst. Akad. Nauk SSSR* 68:3–18 (in Russian).

Carroll, R. L., and Lindsay, W. (1985). Cranial anatomy of the primitive reptile *Procolophon*. *Can. J. Earth Sci.* 22:1571–1587.

Case, E. C. (1911). A revision of the Cotylosauria of North America. *Carnegie Inst. Washington Publ.* 145:1–122.

Cluver, M. A. (1971). The cranial morphology of the dicynodont genus *Lystrosaurus*. *Ann. S. Afr. Mus.* 56:155–274.

Cluver, M. A. (1978). The skeleton of the mammal-like reptile *Cistecephalus* with evidence for a fossorial mode of life. *Ann. S. Afr. Mus.* 76:213–246.

Cluver, M. A., and King, G. M. (1983). A reassessment of the relationships of Permian Dicynodontia (Reptilia, Therapsida) and a new classification of dicynodonts. *Ann. S. Afr. Mus.* 91:195–273.

Crompton, A. W. (1972). Postcanine occlusion in cynodonts and tritylodontids. *Bull. Brit. Mus. (Nat. Hist.), Geol.* 21:27–71.

Crompton, A. W., and Hotton, N. III. (1967). Functional morphology of the masticatory apparatus of two dicynodonts (Reptilia: Therapsida). *Postilla* 109:1–51.

Das, I. (1996). Folivory and seasonal changes in diet in *Rana hexadactyla* (Anura: Ranidae). *J. Zool., Lond.* 238:785–794.

Dearing, M. D. (1993). An alimentary specialization for herbivory in the tropical whiptail lizard *Cnemidophorus murinus*. *J. Herpet.* 27:111–114.

DeMar, R., and Bolt, J. R. (1981). Dentitional organization and function in a Triassic reptile. *J. Paleont.* 55:967–984.

Dilkes, D. W. (1995). The rhynchosaur *Howesia browni* from the Lower Triassic of South Africa. *Palaeontology* 38:665–685.

Dilkes, D. W. (1998). The Early Triassic rhynchosaur *Mesosuchus browni* and the interrelationships of basal archosauromorph reptiles. *Phil. Trans. R. Soc. Lond. B* 353:501–541.

Dodick, J. T., and Modesto, S. P. (1995). The cranial anatomy of the captorhinid reptile *Labidosaurikos meachami* from the Lower Permian of Oklahoma. *Palaeontology* 38:687–711.

Eaton, T. H., Jr. (1964). A captorhinomorph predator and its prey (Cotylosauria). *Amer. Mus. Novit.* 2169:1–3.

Efremov, J. A. (1940). *Ulemosaurus svijagensis* Riab. – ein Dinocephale aus den Ablagerungen des Perm der UdSSR. *Nova Acta Leopold., N. F.* 9(59):155–205.

Farlow, J. O. (1987). Speculations about the diet and digestive physiology of herbivorous dinosaurs. *Paleobiology* 13:60–72.

Gow, C. E. (1978). The advent of herbivory in certain reptilian lineages during the Triassic. *Palaeont. Afr.* 21:133–141.

Gregory, J. T. (1945). Osteology and relationships of *Trilophosaurus*. *Univ. Texas Publ.* no. 4401:273–359.

Gregory, W. K. (1926). The skeleton of *Moschops capensis* Broom, a dinocephalian reptile from the Permian of South Africa. *Bull. Amer. Mus. Nat. Hist.* 56:179–251.

Grine, F. E. (1977). Postcanine tooth function and jaw movement in the gomphodont cynodont *Diademodon* (Reptilia: Therapsida). *Palaeont. Afr.* 20:123–135.

Harland, W. B., Armstrong, R. L., Cox, A. V., Craig, L. E., Smith, A.G., and Smith, D. G. (1990). *A Geologic Time Scale 1989*. Cambridge and New York: Cambridge University Press.

Heaton, M. J. (1979). Cranial anatomy of primitive captorhinid reptiles from the Late Pennsylvanian and Early Permian, Oklahoma and Texas. *Oklahoma Geol. Surv. Bull.* 127:1–84.

Heaton, M. J. (1980). The Cotylosauria: a reconsideration of a group of archaic tetrapods. In *The Terrestrial Environment and the Origin of Land Vertebrates*, ed. A. L. Panchen, pp. 497–551. London and New York: Academic Press.

Hopson, J. A. (1971). Postcanine replacement in the gomphodont cynodont *Diademodon*. In *Early Mammals*, ed. D. M. Kermack and K. A. Kermack, pp. 1–21. London: Academic Press.

Hopson, J. A. (1984). Late Triassic traversodont cynodonts from Nova Scotia and southern Africa. *Palaeont. Afr.* 25:181–201.

Hotton, N. III, Olson, E.C., and Beerbower, R. (1997). Amniote origins and the discovery of herbivory. In *Amniote Origins*, ed. S. S. Sumida and K. L. M. Martin, pp. 207–264. San Diego: Academic Press.

Ivakhnenko, M. F., and Tverdokhlebova, G. I. (1987). [Revision of Permian bolosauromorphs from eastern Europe.] *Paleont. Zh.* 1987(2):98–106 (in Russian).

Iverson, J. B. (1982). Adaptations to herbivory in iguanine lizards. In *Iguanas of the World: Their Behavior, Ecology and Conservation*, ed. G. M. Burghardt and A. S. Rand, pp. 77–83. Park Ridge, NJ: Noyes Publications.

Janis, C. M. (1995). Correlations between craniodental morphology and feeding behavior in ungulates: reciprocal illumination between living and fossil taxa. In

Functional Morphology in Vertebrate Paleontology, ed. J. J. Thomason, pp.76–98. Cambridge and New York: Cambridge University Press.

Kemp, T. S. (1982). *Mammal-like Reptiles and the Origin of Mammals.* London and New York: Academic Press.

King, G. (1990). *The Dicynodonts: A Study in Palaeobiology*. London: Chapman and Hall.

King, G. (1996). *Reptiles and Herbivory*. London: Chapman and Hall.

King, G. M., Oelofsen, B. W., and Rubidge, B. S. (1989). The evolution of the dicynodont feeding system. *Zool. J. Linn. Soc.* 96:185–211.

Kitching, J. W. (1977). The distribution of the Karroo vertebrate fauna. *Bernard Price Inst. Palaeont. Res., Mem.* 1:1–131.

Labandeira, C. C. (1997). Insect mouthparts: Ascertaining the paleobiology of insect feeding strategies. *Annu. Rev. Ecol. Syst.* 28:153–193.

Laurin, M., and Reisz, R. R. (1995). A reevaluation of early amniote phylogeny. *Zool. J. Linn. Soc.* 113:165–223.

Lee, M. S. Y. (1996). Correlated progression and the origin of turtles. *Nature* 379:812–815.

Lee, M. S. Y. (1997). Pareiasaur phylogeny and the origin of turtles. *Zool. J. Linn. Soc.* 120:197–280.

Lee, M. S. Y., and Spencer, P. S. (1997). Crown clades, key characters and taxonomic stability: When is an amniote not an amniote? In *Amniote Origins*, ed. S. S. Sumida and K. L. M. Martin, pp. 61–84. San Diego: Academic Press.

Li J. and Cheng Z. (1995). [The first discovery of bolosaurs from the Upper Permian of China.] *Vert. PalAsiat.* 33:17–23 (in Chinese with English summary).

Maynard Smith, J., and Savage, R. J. G. (1959). The mechanics of mammalian jaws. *School Sci. Rev.* 141:289–301.

Modesto, S. P. (1992). Did herbivory foster early amniote diversification? *J. Vert. Paleont.* 12 (Suppl. to 3):44A.

Modesto, S. P. (1994). The Lower Permian synapsid *Glaucosaurus* from Texas. *Palaeontology* 37:51–60.

Modesto, S. P. (1995). The skull of the herbivorous synapsid *Edaphosaurus boanerges* from the Lower Permian of Texas. *Palaeontology* 38:213–239.

Modesto, S. P., and Reisz, R. R. (1990). A new skeleton of *Ianthasaurus hardestii*, a primitive edaphosaur (Synapsida: Pelycosauria) from the Upper Pennsylvanian of Kansas. *Can. J. Earth Sci.* 27:834–844.

Olson, E. C. (1947). The family Diadectidae and its bearing on the classification of reptiles. *Fieldiana, Geol.* 11:1–53.

Olson, E. C. (1961). The food chain and the origin of mammals. In *International Colloquium on the Evolution of Lower and Non-Specialized Mammals*, ed. G. Vanderbroek, pp. 97–116. Brussels: Koninklijke Vlaamse Akademie voor Wetenschappen, Letteren en Schone Kunsten van België.

Olson, E. C. (1966). Community evolution and the origin of mammals. *Ecology* 47:291–308.

Olson, E. C. (1968). The family Caseidae. *Fieldiana, Geol.* 17:225–349.

Parra, R. (1978). Comparison of foregut and hindgut fermentation in herbivores. In *The Ecology of Arboreal Folivores*, ed. G. G. Montgomery, pp. 205–229. Washington, DC: Smithsonian Institution Press.

Parrington, F. R. (1946). On the cranial anatomy of cynodonts. *Proc. Zool. Soc. Lond.* 116 (II):181–197.

Reisz, R. R. (1986). Pelycosauria. In *Handbuch der Paläoherpetologie, Teil* 17A, ed. P. Wellnhofer, pp. 1–102. Stuttgart: Gustav Fischer Verlag.

Reisz, R. R. (1997). The origin and early evolutionary history of amniotes. *Trends Ecol. Evol.* 12:218–222.
Rensberger, J. M. (1986). Early chewing mechanisms in mammalian herbivores. *Paleobiology* 12:474–494.
Ricqlès, A. de, and Taquet, P. (1982). La faune de vertébrés du Permien Supérieur du Niger. I. Le captorhinomorphe *Moradisaurus grandis* (Reptilia, Cotylosauria) – Le crâne. *Ann. Paléont.* 68:33–106.
Romer, A. S. (1967). The Chañares (Argentina) Triassic reptile fauna. III. Two new gomphodonts, *Massetognathus pascuali* and *M. teruggii*. *Breviora* 264:1–25.
Romer, A. S., and Price, L. I. (1940). Review of the Pelycosauria. *Geol. Soc. Amer. Spec. Pap.* 28:1–538.
Rubidge, B. S., and Hopson, J. A. (1996). A primitive anomodont therapsid from the base of the Beaufort Group (Upper Permian) of South Africa. *Zool. J. Linn. Soc.* 117:115–139.
Rybczynski, N. (1996). *Cranial Morphology and Phylogenetic Significance of* Suminia getmanovi, *a Late Permian Anomodont from Russia*. M.Sc. thesis, University of Toronto.
Schiek, J. O., and Millar, J. S. (1985). Alimentary tract measurements as indicators of diet in small mammals. *Mammalia* 49:93–104.
Scott, A. C., and Taylor, T. N. (1983). Plant/animal interactions during the Upper Carboniferous. *Bot. Rev.* 49:259–307.
Smith, R. M. H. (1987). Helical burrow casts of therapsid origin from the Beaufort Group (Permian) of South Africa. *Palaeogeogr., Palaeoclimat., Palaeoecol.* 60:155–170.
Stovall, J. W., Price, L. I., and Romer, A. S. (1966). The postcranial skeleton of the giant Permian pelycosaur *Cotylorhynchus romeri*. *Bull. Mus. Comp. Zool., Harvard Univ.* 135:1–30.
Sues, H.-D., and Baird, D. (1998). Procolophonidae (Amniota: Parareptilia) from the Wolfville Formation (Upper Triassic: Carnian) of Nova Scotia, Canada. *J. Vert. Paleont.* 18:525–532.
Sues, H.-D., and Reisz, R. R. (1998). Origin and early evolution of herbivory in terrestrial tetrapods. *Trends Ecol. Evol.* 18:141–145.
Sues, H.-D., Olsen, P. E., Scott, D. M., and Spencer, P. S. (2000). Cranial osteology of *Hypsognathus fenneri*, a latest Triassic procolophonid reptile from the Newark Supergroup of eastern North America. *J. Vert. Paleont.* 20.
Throckmorton, G. S. (1976). Oral food processing in two herbivorous lizards, *Iguana iguana* (Iguanidae) and *Uromastix aegyptius* (Agamidae). *J. Morph.* 148:363–390.
Troyer, K. (1982). Transfer of fermentative microbes between generations in a herbivorous lizard. *Science* 216:540–542.
Turnbull, W. D. (1970). Mammalian masticatory apparatus. *Fieldiana, Geol.* 18:149–356.
Watson, D. M. S. (1954). On *Bolosaurus* and the origin and classification of reptiles. *Bull. Mus. Comp. Zool., Harvard Univ.* 111:297–449.
Weishampel, D. B., and Norman, D. B. (1989). Vertebrate herbivory in the Mesozoic: jaws, plants, and evolutionary metrics. In *Paleobiology of the Dinosaurs*, ed. J. O. Farlow, Jr., pp. 87–100. *Geol. Soc. Amer. Special Paper* 238.
Williston, S. W. (1911). A new family of reptiles from the Permian of New Mexico. *Amer. J. Sci.* 31:378–398.
Wing, S. L., and Sues, H.-D. (rapporteurs) (1992). Mesozoic and early Cenozoic terrestrial ecosystems. In *Terrestrial Ecosystems through Time*, ed. A. K. Behrensmeyer *et al.*, pp. 327–416. Chicago: University of Chicago Press.

PAUL M. BARRETT

3

Prosauropod dinosaurs and iguanas: speculations on the diets of extinct reptiles

Introduction

A thorough knowledge of extant taxa is required when attempting paleobiological interpretations of extinct animals (Lauder 1995; Witmer 1995). Skeletal material, though informative, can provide only a limited amount of information on soft-tissue and behavioral characteristics. It is not possible to identify with certainty the actual diet of any extinct animal because we cannot observe the animal feeding. Fossilized gut contents (enterolites) and feces (coprolites) may provide some dietary information, but convincing enterolites are rare and coprolites are almost never associated with the animal that produced them. In some cases, direct comparison of the feeding adaptations of extinct taxa with closely related extant forms ('Extant Phylogenetic Bracket' approach *sensu* Witmer 1995) is possible, allowing these interpretations to be made with some confidence (e.g., between extinct and extant mammals). Some groups, however, have no present-day representatives for relevant comparison. For example, study of extant birds and crocodilians provides only limited insight into the paleobiology of non-avian dinosaurs. In cases where the phylogenetic approach to assessing function in extinct organisms is not feasible, there is a tendency to rely on ahistorical biomechanical modeling (Weishampel 1995). Such modeling is often based on the identification and study of suitable extant analogues. The choice of extant taxa for comparison, however, is often limited, and many studies are forced to utilize animal models that may be distantly related to the extinct taxon under consideration. This chapter deals with one such model – the relationship between tooth form and diet in herbivorous reptiles.

Ever since the original description of *Iguanodon* (Mantell 1825), the structure of the more posterior teeth of extant iguanine lizards has been used as a paradigm for deducing the diet of extinct reptiles. Mantell noticed that the teeth of both *Iguanodon* and *Iguana* had crowns that were mesiodistally expanded and coarsely serrated along their mesial and distal edges. As extant iguanines were thought to be exclusively herbivorous, Cuvier (in Mantell 1825) reasoned, on the basis of tooth form, that *Iguanodon* was a gigantic herbivorous reptile. Since then, all extinct reptiles with this characteristic type of tooth structure have been interpreted as herbivores (e.g., Galton 1985a, 1986; Crompton and Attridge 1986).

In this chapter, special attention will be paid to the comparisons that have been made between the dentitions of prosauropod dinosaurs and extant iguanine lizards. The diet of prosauropods has been the subject of considerable controversy, not because authors have disagreed over the dental features present in these animals, but because the interpretations based on these features have differed considerably. Prosauropods were a group of medium-sized to large saurischian dinosaurs with a worldwide distribution during the Late Triassic and Early Jurassic (Galton 1990). This group is currently thought to be monophyletic (Sereno 1989; Galton 1990; Upchurch 1993; Gauffre 1995), although some authors have argued that prosauropods are paraphyletic with respect to sauropods (Gauthier 1986; Benton 1990). Prosauropods were taxonomically diverse and numerically abundant, and include well-known forms like *Plateosaurus* and *Massospondylus*.

Institutional abbreviations used in this chapter are: BMNH, The Natural History Museum, London; BP/1, Bernard Price Institute for Palaeontological Research, University of the Witswatersrand, Johannesburg; FMNH, Field Museum, Chicago; IVPP, Institute of Vertebrate Paleontology and Paleoanthropology, Academia Sinica, Beijing; MCZ, Museum of Comparative Zoology, Harvard University; UCMZ, University Museum of Zoology, University of Cambridge.

Previous work on prosauropod diets

Historical review

The diet of prosauropods has prompted considerable interest among paleontologists. Almost every conceivable mode of life has been postulated for prosauropods, ranging from active carnivores (Swinton 1934) to exclusively herbivorous forms (Galton 1984a, 1985a, 1986). This diversity of opinion seems, in some regard, to follow the prevailing views on prosauropod classification. Earlier authors tended to place the Prosauropoda within the Theropoda, perhaps predisposing these authors to view prosauropods as carnivores (Huene 1932; Swinton 1934; Lull 1953). Swinton (1934:58) regarded *Plateosaurus* as '[an] enemy of unusual power and terrifying aspect' based on the large size of the animal, the structure of the large manus claw and the serrated tooth form; he further suggested that these features may have been adaptations for piscivory. Lull (1953:118) concluded that *Anchisaurus* ('*Yaleosaurus*') was 'an alert, active dinosaur, preying upon the smaller vertebrates of his generation, as the powerful claws and well developed teeth imply.' Huene (1932) also commented on the prosauropod dentition, describing that of *Ammosaurus* as 'carnivorous.' More recent phylogenetic hypotheses, which allied the prosauropods with the sauropods in the Sauropodomorpha (Charig, Attridge and Crompton 1965; Romer 1966), opened the way for reinterpretations of prosauropod feeding behaviour and brought the prosauropods into line with the undoubtedly herbivorous sauropods. Colbert (1962:70–71) regarded *Plateosaurus* as exclusively herbivorous, feeding on 'soft vegetation', while consideration of the dentition of *Anchisaurus* ('*Yaleosaurus*') led him to the conclusion that '*Yaleosaurus* . . . was turning from a diet of animals, and depending for much of its nourishment on the plants of those distant Triassic days.' However, the common association of prosauropod skeletons with recurved, finely serrated teeth, like those of carnivorous archosaurs, led some authors (Charig *et al.* 1965; Romer 1966) to suggest that some prosauropods (melanorosaurids and *Teratosaurus* in particular) were carnivorous. These associations are now regarded as fortuitous; the recurved teeth were probably shed by carnivorous archosaurs feeding on prosauropod carcasses (Galton 1984a, 1985a, 1986). Furthermore, *Teratosaurus* has been shown to represent a poposaurid archosaur (Galton 1985b), rather than a carnivorous prosauropod dinosaur. Charig *et al.* (1965) argued that there was no reason to try to 'shoe-horn' all prosauropods into a particular ecological category, and speculated that prosauropods covered a wide range of ecological preferences. They proposed that the group may have included herbivorous, carnivorous, and omnivorous representatives, but this idea was largely ignored by subsequent authors. However, despite this diversity of opinion, a consensus seems to have been reached sometime in the early 1970s that prosauropods were primarily herbivorous (e.g., Galton 1973, 1976).

Following a detailed study of the southern African prosauropod *Mas-*

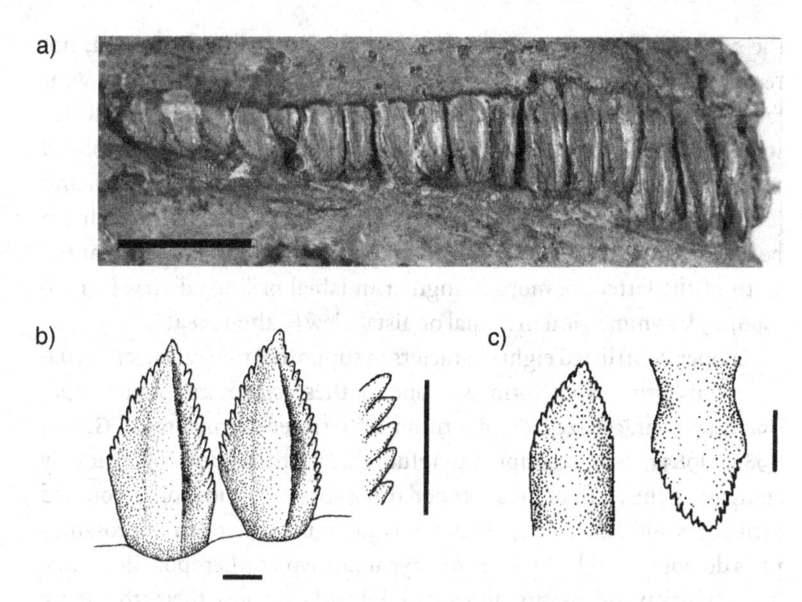

Figure 3.1. Prosauropod teeth in labial view. a) *Massospondylus* (BP/1/4376). Scale bar equals 10 mm. b) *Thecodontosaurus* (inset with detail of the cutting edge showing denticles). Scale bars each equal 1 mm. c) *Sellosaurus*. Scale bar equals 5 mm. (b and c from Galton 1985a; courtesy and copyright of *Lethaia*.)

sospondylus, Cooper (1981) challenged this view and suggested that prosauropods were scavenger-predators. Kermack (1984) also cast doubts on prosauropod herbivory, concluding that *Thecodontosaurus* was an omnivore. Countering these views, Galton (1984a, 1985a, 1986) argued forcefully in support of prosauropod herbivory. The main arguments in this debate are reviewed below.

Scavenger-predators or herbivores?

The single most important feature employed in deducing prosauropod diets, used by all of the authors mentioned above, has been the form of the teeth. The following description of the prosauropod dentition is based on the detailed observations of Galton (1984a, 1985a, 1986), supplemented by personal observations on a wide range of prosauropod taxa. In general, prosauropod maxillary and dentary teeth have crowns that are mesiodistally expanded (clearly separating the crown from the root), taller than wide in labial or lingual view, and symmetrical in mesial or distal view (Figure 3.1). The tooth crowns are coarsely serrated along their mesial and distal edges with the serrations projecting at an angle of about 45° to the edge of the tooth crown. They are labiolingually compressed, not recurved, and lack cingula. Wear is usually absent, but small apical wear facets have been reported in *Massospondylus* (Gow, Kitching and Raath 1990), and high-angle mesial and distal wear facets are present on isolated teeth referred to *Plateosaurus* (Cuny and Ramboer 1991) and *Yunnanosaurus* (Galton 1985a; see below). Prosauropod maxillary and dentary teeth can be distinguished from those of contemporary ornithischians in that the teeth of the latter are more triangular in labial or lingual view (Sereno 1986) and asymmetrical in mesial or distal view (Galton 1984b).

Cooper (1981) listed eight characters in support of the scavenger-predator hypothesis. Tooth form was one of these characters, and Cooper (1981:814) characterized the dentition as 'serrated, carnivorous.' Galton (1984a, 1985a, 1986) attempted to refute this particular interpretation by comparing the teeth of prosauropod dinosaurs with those of undoubted carnivores and herbivores. He demonstrated that the teeth of prosauropods do not resemble those of archetypal carnivores (theropod dinosaurs, 'thecodontian' archosaurs, and varanid lizards) because their crowns are not recurved, are more compressed labiolingually and mesiodistally expanded, and, perhaps most importantly, differ in the form of their serrations. The teeth of carnivorous forms possess very fine serrations, which project from the crown at an angle of around 90° to the edge of the tooth (Figure 3.2a,b). In contrast, the serrations on prosauropod teeth are coarser (i.e., larger relative to crown length), less numerous, and project at an angle of around 45° from the edge of the tooth crown (see above and Figure 3.1). Galton noted that prosauropod teeth are very similar to those of extant iguanine lizards (notably *Iguana*; Figure 3.3a). Previous work on iguanine diets (Hotton 1955; Montanucci 1968) appeared to indicate that *Iguana* was exclusively herbivorous, and Galton concluded that prosauropods were also herbivores, using their teeth to cut and tear plant material in the same way as iguanine lizards (Throckmorton 1976). Galton's views have subsequently been accepted by most paleontologists.

Kermack (1984) noted that the 'leaf-shaped' teeth of *Thecodontosaurus* indicated an herbivorous diet, but noted the absence of other adaptations to herbivory, such as the presence of a ventrally displaced jaw joint. Unable to reconcile the simple 'carnivorous' jaw action of *Thecodontosaurus* with the 'herbivore-like' tooth structure, she advocated a compromise, proposing that *Thecodontosaurus* was omnivorous. Kermack (1984:110) envisioned *Thecodontosaurus* 'eating largely soft vegetable matter but supplementing

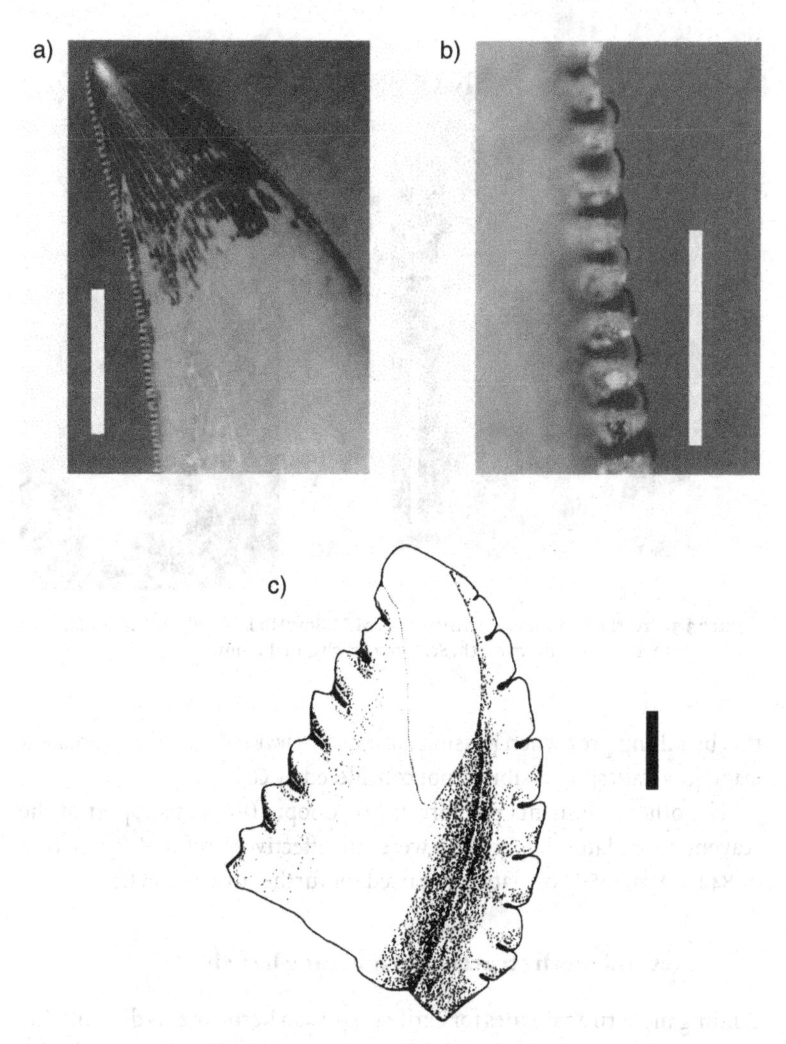

Figure 3.2. Tooth crowns of theropod dinosaurs. a,b) Tooth crown of indeterminate theropod (BMNH R5226) showing the recurvature and fine serrations characteristic of carnivorous dentitions. a) ?Labial view of tooth crown. Scale bar equals 10 mm. b) Detail of the serrations on the distal cutting edge of the tooth. Scale bar equals 2 mm. c) Premaxillary tooth of *Troodon formosus* in lingual view (from Currie 1987; courtesy and copyright of *Journal of Vertebrate Paleontology*). Scale bar equals 1 mm.

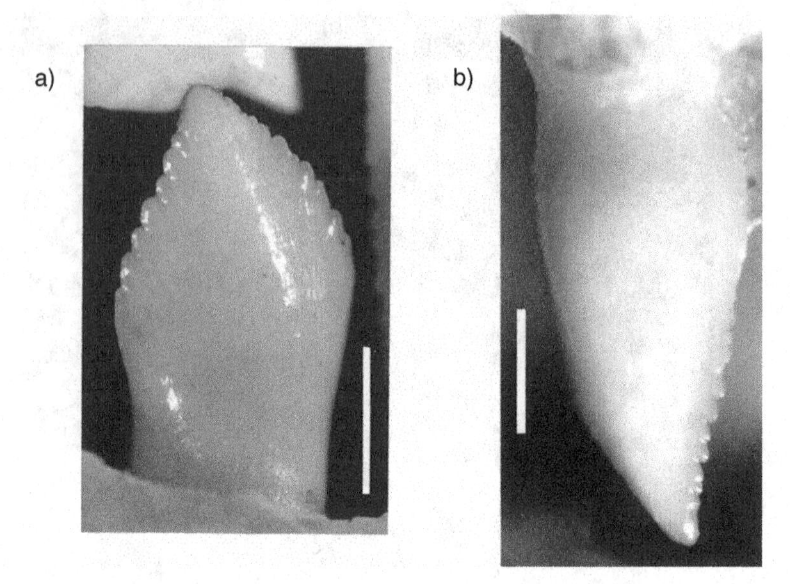

Figure 3.3. Teeth of *Iguana ?delicatissima* (UCMZ R8917) in labial view. a) Maxillary tooth. b) Premaxillary tooth. Scale bars each equal 1 mm.

this by killing prey when possible, *in a similar way to* Iguana *or* Uromastix *today*' [my italics], a possibility not considered by Galton.

The other arguments put forward by Cooper (1981) in support of the scavenger-predator hypothesis were all effectively refuted by Galton (1984a, 1985a, 1986), obviating the need for further discussion here.

Diet and tooth structure in iguanine lizards

Finding modern analogues for extinct reptilian herbivores is difficult due to the paucity of present-day herbivorous reptiles. Today, reptilian herbivores are represented by a small number of lizards and turtles (King 1996). Turtles lack teeth and crop vegetation using a self-sharpening keratinous beak, limiting study of the relationship between tooth form and diet to herbivorous lizards, many of which are referable to the Iguanidae. The majority of iguanids are insectivorous whereas members of the Iguaninae (e.g., *Iguana, Ctenosaura*) are thought to be exclusively herbivorous (Hotton 1955; Montanucci 1968). Several authors have argued for a close relationship between tooth form and diet in iguanid lizards. Cuspidation is poorly developed in insectivores whereas herbivores possess cuspidate

Table 3.1. *Food items taken by free-living and captive iguanine lizards*

Iguana iguana	*Dipsosaurus dorsalis*	*Uromastix* spp.
Small mammals	Small mammals (rarely)	Small mammals
Baby mice	Insects	Insects
Birds	Eggs (rarely)	Fruit
Insects	Chopped meat (rarely)	Vegetables
Dog food	Vegetables	
Chopped meat	Fruit	
Fruit	Leaves, shoots	
Vegetables		
Dandelion flowers		
Mealworms		
Eggs (rarely)		
Leaves, shoots		

Sources: Hotton (1955); Montanucci (1968); Wallach and Boever (1983); Frey (1986).

or serrated teeth (Hotton 1955; Montanucci 1968). Interestingly, a number of herbivorous fishes possess coarsely serrated teeth very similar to those of iguanines. The surgeonfish *Acanthurus nigrofuscus* uses these teeth to tear or incise algae from rocks, and *Ctenochaetus striatus* employs them to rake particles from algal stands (Purcell and Bellwood 1993).

Contrary to many published accounts, 'herbivorous' iguanines are not exclusively herbivorous. Captive iguanines are known to eat a wide variety of foods, ranging from fruit to dog food (Frye 1974 [cited in Wallach and Boever 1983]; Frey 1986; Table 3.1), and show severe growth disturbances if meat is excluded from their diet (J. Pearson, pers. comm.). These observations indicate that free-living iguanines, of necessity, must include animal protein in their diets to avoid various vitamin-deficiency disorders. Previous authors often deduced the diet of free-living iguanines from dissection and inspection of stomach contents. These dissections have shown a preponderance of plant material in the gut with rare animal remains such as insect cuticle. However, these dissections are likely to be misleading for two reasons (E. N. Arnold, pers. comm.). First, plant material is very resistant to digestion and has a longer passage time through the gut than meat. Second, plant material is nutritionally poorer, on a weight for weight basis, than animal protein. Therefore, a greater volume of vegetation must be consumed in order to provide the same amount of nutrients as a much smaller quantity of meat. Other factors also need to be taken into consideration. The diet of iguanines has been shown to change during ontogeny (Montanucci 1968), and there has been a general disregard of the selective pressures exerted on juveniles by the relationship between diet and dentition, as most authors have concentrated on adults in their studies. Furthermore, little if any attention has been paid to seasonal variation in the abundance of various food items and the impact of these changes. Is it possible that, depending on environmental conditions, these animals can spend extended periods as strict herbivores or carnivores? More field data are needed on the dietary preferences of free-living iguanines, but anecdotal evidence shows them preying on small rodents, eggs, and hatchling birds in addition to their more generally accepted herbivorous diet (Table 3.1 and references cited therein). Such observations are likely to provide additional surprises: for example, giant tortoises on Aldabara are known to feed on dead tortoises and human feces (E. N. Arnold, pers. comm.), whereas mammalian herbivores like horses, red deer, sheep and cattle have been observed eating nesting seabirds, chicks, and discarded hamburgers (Furness 1988a, 1988b, 1989; S. Finney, pers. comm.). It seems more reasonable to interpret *Iguana* and other 'herbivorous' iguanine lizards as opportunistic or facultatively omnivorous rather than strictly herbivorous.

The serrated tooth structure of 'herbivorous' iguanines may be an adaptation to herbivory in the context of continued carnivory or omnivory, rather than an *a priori* adaptation to herbivory (a conclusion reached independently by E. N. Arnold [pers. comm.]). This hypothesis was tested by using a comparative approach to iguanine diets (Figure 3.4). Data on iguanine diets were obtained from a number of sources (Hotton 1955; Montanucci 1968; Wallach and Boever 1983; Frey 1986). Unfortunately, the only available data are either qualitative or anecdotal; good quantitative information is lacking. These data were plotted onto a cladogram of iguanine interrelationships (de Queiroz 1987), together with the tooth form of the taxa under consideration. *Iguana* and *Cyclura* are the most derived iguanines (de Queiroz 1987) and are also the taxa in which vegetation makes up the greatest part of the diet (Figure 3.4). Less derived iguanines, such as *Ctenosaura*, are known to eat less vegetation (Hotton 1955; Montanucci 1968). *Iguana* possesses the most cuspidate teeth whereas other iguanines, which eat more insects, have teeth with fewer cusps. Thus there is a correlation between diet, tooth structure, and phylogeny. However, it is notable that significant quantities of animal prey are still

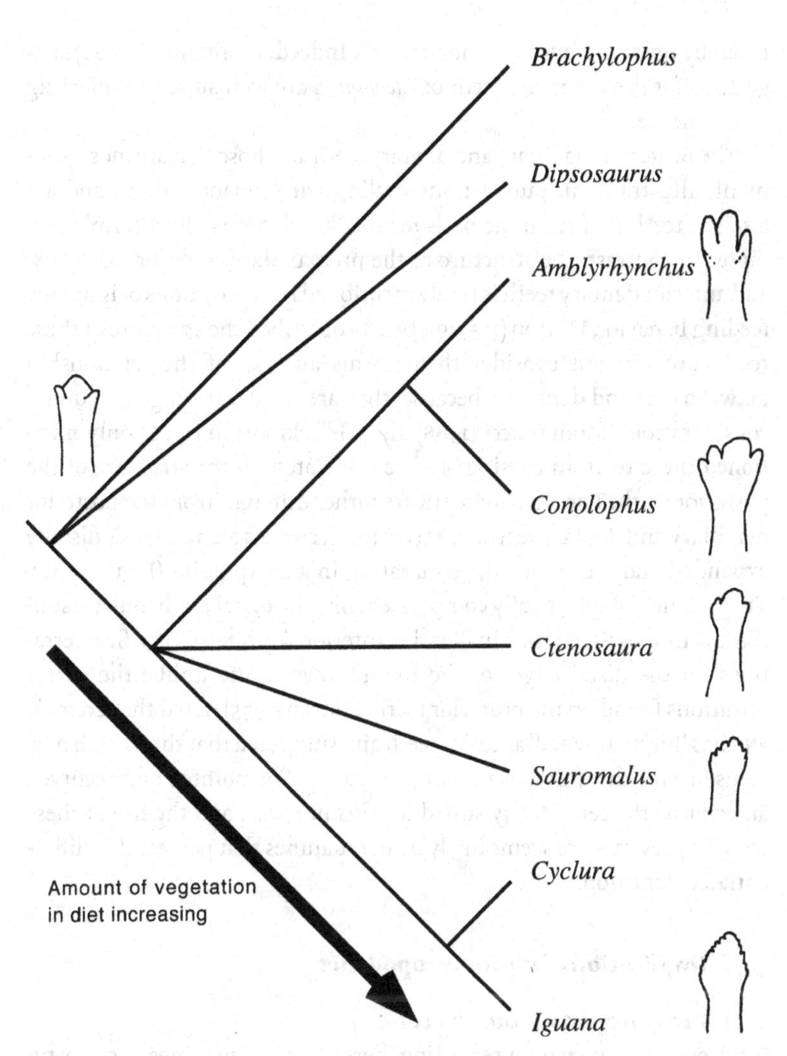

Figure 3.4. Correlation between iguanine phylogeny, tooth form, and diet. The tooth of a basal, insectivorous iguanid (*Basiliscus*) is shown at the left. Phylogeny based on de Queiroz (1987), iguanine tooth form based on de Queiroz (1987) and Montanucci (1968), and diet based on various sources including Hotton (1955), Montanucci (1968), and Frey (1986). Drawings are not to scale.

taken by iguanines with cuspidate teeth. Indeed, Montanucci (1968) suggested that the cuspidate teeth of *Ctenosaura* are well-suited for piercing insect cuticle.

The posterior maxillary and dentary teeth are those that are most commonly illustrated in publications dealing with iguanine diets, and are also the teeth that form the basis for the 'herbivorous tooth-form' paradigm. In contrast, the structure of the premaxillary, anterior maxillary, and anterior dentary teeth is rarely mentioned in descriptions of iguanine feeding behavior. Hotton (1955:91) briefly described the structure of these teeth, but did not consider them in his analysis of the relationship between diet and dentition because 'they are similar in all genera under consideration.' Montanucci (1968) and Throckmorton (1976) only mentioned these teeth in passing, and concentrated on the structure of the posterior teeth. The anterior teeth are rather different from the posterior maxillary and dentary teeth. Anterior tooth crowns are not mesiodistally expanded and are recurved, terminating in a sharp point (Figure 3.3b). They are not labiolingually compressed, like the distal teeth, but are subconical in cross-section. Finally, the anterior teeth bear very fine serrations on the distal edges of the tooth crown, quite unlike the coarse serrations found on the posterior teeth. Hotton (1955) noted that *Crotaphytus* has 'high-crowned' anterior teeth and suggested that these teeth may be useful for the quick seizure of prey items. The pointed and recurved anterior teeth seem ideally suited for this purpose, and the use of these teeth in prey capture seems likely in all iguanines that possess this differentiated dentition.

Implications for prosauropod diet

Prosauropods as omnivores

If the evidence in favor of regarding 'herbivorous' iguanines as opportunistic or facultative omnivores is accepted, the utility of iguanine tooth form as a paradigm for inferring the diet of extinct reptiles needs to be reassessed. The possession of iguanine-like teeth can no longer be used as a reliable indicator of herbivory *sensu stricto*. Indeed, if we are to use iguanine lizards as a paradigm for establishing diet, we should conclude that other animals possessing this type of dentition were also omnivorous.

The striking similarities in the dentitions of prosauropod dinosaurs and iguanine lizards indicate that it is reasonable to regard prosauropods as opportunistic or facultative omnivores, a possibility mentioned by several other authors (Kermack 1984; Gow, Kitching, and Raath 1990; Zhang and Yang 1994). Prosauropods have recurved premaxillary teeth and anterior maxillary teeth that sometimes bear very fine serrations along their distal edges (e.g. *Massospondylus* [Gow *et al.* 1990; MCZ 8893], *Jingshanosaurus* [Zhang and Yang 1994]) (Figure 3.5). These teeth could have functioned in a similar way to the anterior teeth of iguanines in the capture of small prey (e.g., reptiles, mammals, invertebrates). An undescribed anchisaurid skeleton from the Lower Jurassic McCoy Brook Formation of Nova Scotia provides some support for this hypothesis. This specimen was found with an *in situ* gastric mill that contained a badly worn maxilla of the sphenodontid *Clevosaurus* (Shubin, Olsen, and Sues 1994; H.-D. Sues, pers. comm.).

Several other features of prosauropods support the contention that they may have had more varied diets than usually supposed. As Cooper (1981) noted, there is no reason to suggest that the large manual ungual of prosauropods could not have been used in dismembering carrion in addition to its other functions. Furthermore, supposedly 'herbivorous' features including the ventrally displaced jaw joint, precise occlusion, and a well-developed coronoid eminence, are poorly developed or absent in many prosauropod dinosaurs (Galton 1976; Barrett 1998). It is possible that different prosauropod taxa had different diets, as originally suggested by Charig *et al.* (1965).

The degree to which prosauropods exploited animal material probably depended on environmental factors, age, and ecological opportunities. Possible environmental and ontogenetic factors are discussed below.

Environmental factors

Prosauropod skeletal remains have been recovered from a variety of paleoenvironmental settings, ranging from lacustrine deposits with abundant plant material to deposits formed under arid conditions with ephemeral water sources and sparse vegetation (see Table 3.2). The majority of prosauropod fossils have been found in sedimentary rocks from depositional settings that are best described as 'semi-arid' or 'seasonally wet.' These environments were typically floodplains or alluvial fans with ephemeral surface water (braided streams and playa lakes) and occasional permanent water bodies. In some prosauropod-bearing strata, there is good evidence of rapid climatic fluctuations (e.g., Fleming Fjord Formation of East Greenland [Jenkins *et al.* 1994]). The majority of prosauropod

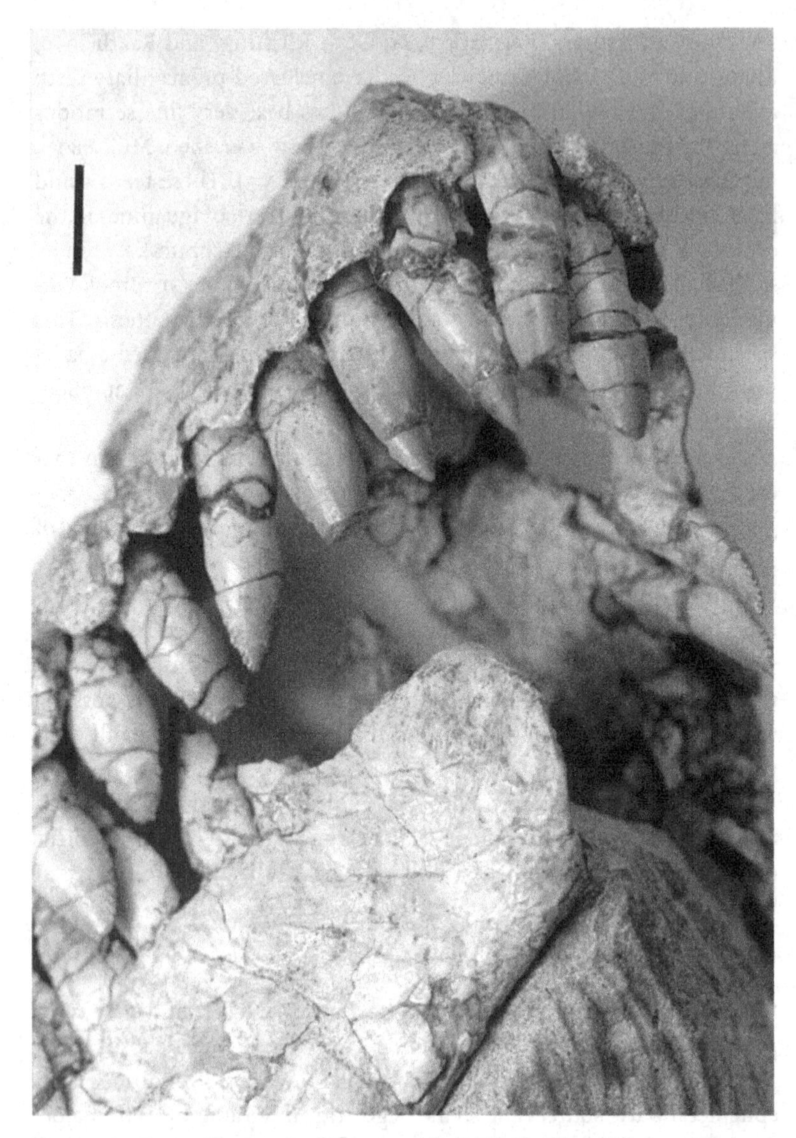

Figure 3.5. Premaxillary teeth of *Massospondylus* (MCZ 8893) in labial view. Scale bar equals 10 mm.

Table 3.2. *Paleoenvironments of some of the principal prosauropod-bearing formations*

Formation/locality	Taxa present	Age	Paleoenvironment	Source(s)
Portland Formation, Newark Supergroup, Connecticut, USA	*Anchisaurus* *Ammosaurus*	Sinemurian–Pliensbachian	The lower part of the formation consists of several lacustrine cycles. Wet periods alternate with much drier periods. The upper part is characterized by ephemeral streams, suggestive of a much drier climate	Smoot (1991)
Navajo Sandstone, Arizona, USA	*Ammosaurus*	Pliensbachian–Toarcian	Vertebrate remains are found in interdune deposits. These interdunes were relatively mesic with respect to the surrounding dune systems	Winkler *et al.* (1991)
Kayenta Formation, Arizona, USA	*Massospondylus*	Hettangian	Broad, well-drained floodplain with abundant surface water surrounded by sand dunes and highlands	Harshbarger, Repenning and Irwin (1957); Colbert (1981); Clark and Fastovsky (1986)
McCoy Brook Formation, Nova Scotia, Canada	cf. *Ammosaurus*	Hettangian	Seasonally wet	H.-D. Sues (pers. comm.)
Fleming Fjord Formation, East Greenland	*Plateosaurus*	Norian	Cyclic deposition in a shallow or ephemeral lake	Jenkins *et al.* (1994)
Lower Elliot Formation, South Africa	*Euskelosaurus* *Blikanasaurus* *Melanorosaurus*	late Carnian or early Norian	Semi-arid climate with meandering channel and floodplain deposits. Perennial and ephemeral water sources	Smith *et al.* (1993)
Upper Elliot Formation, South Africa; Clarens Formation, South Africa; Forest Sandstone Formation, Zimbabwe	*Massospondylus*	Hettangian – ?Sinemurian Sinemurian Hettangian – Sinemurian	Semi-arid to arid climate with flood basin deposition and later flood-fans and dunes. Braided streams, playa and dunes present	Smith *et al.* (1993)

Table 3.2 (*cont.*)

Formation/locality	Taxa present	Age	Paleoenvironment	Source(s)
Knollenmergel, Germany	*Plateosaurus*	late Norian	Climate semi-arid with periodic heavy rains. The sediments were deposited on a broad floodplain	Sander (1992)
Fissure-fillings, Wales	*Thecodontosaurus*	?Norian	Climate semi-arid with infrequent heavy rains	Robinson (1957); S. E. Evans (pers. comm.)
Los Colorados Formation, Argentina	*Riojasaurus* *Coloradisaurus*	Norian	Warm, humid, lacustrine deposition	Bonaparte (1972)
Lower Lufeng Formation, Yunnan, China	*Lufengosaurus* *Yunnanosaurus* *Jingshanosaurus*	Hettangian–Pliensbachian	Lacustine deposits indicating a warm, wet environment with abundant vegetation	Zhang and Yang (1994); Z. Luo (pers. comm.)

paleoenvironments would probably have had a sparse cover of vegetation for much of the time, though there may have been more pluvial periods, either seasonally or due to longer-term climatic variations, during which times more vegetation would have been available.

Unlike the semi-arid environments of today (e.g., savannah, pampas), which can support large (albeit migratory) populations of mammalian herbivores, the environments inhabited by prosauropod dinosaurs lacked the grasses that represent the bulk of the primary production in such regions today. The vegetation available to prosauropods is likely to have been xerophytic or confined to areas where water was permanently available either on or below the surface (Gow *et al.* 1990). It seems unlikely that prosauropods could have maintained an exclusively herbivorous diet under such environmental conditions, particularly given the high biomass of many prosauropod communities (Galton 1985a), and the ability to exploit and utilize animal prey or carrion would have been advantageous. Dietary preferences may have shifted with changes in climate. For example, more animal food may have been taken during dry periods and more vegetation taken when the climatic conditions were wetter. Interestingly, the prosauropod with the most 'herbivore-like' teeth (*Yunnanosaurus*) comes from one of the few environments in which water availability was probably not limiting (Table 3.2).

Ontogenetic factors

Another possibility is an ontogenetic shift in diet. Juvenile iguanines have been reported to take a higher proportion of animal protein than adults. In *Ctenosaura*, the juveniles have a larger number of cusps on their teeth than adults, but eat less vegetation (Montanucci 1968). This observation contradicts the view that higher levels of tooth cuspidation are an *a priori* adaptation to herbivory because adults, which take more vegetation in their diets, would be expected to have more serrations on their teeth than juveniles. However, fewer cusps on the teeth of adults may simply reflect the longer functional life of the tooth. More work is needed in this area. Prosauropods may also have displayed an ontogenetic shift in diet, with juveniles taking more animal prey than adults. However, no significant changes in tooth structure could be identified in a growth series of *Massospondylus* (BP/1/4376, BP/1/4779, BP/1/5241). The teeth of the juvenile specimen (BP/1/4376) possess one or two fewer denticles per tooth crown than the other, larger, specimens, but the increase in the number of denticles on adult teeth probably reflects the larger size of adult teeth rather than a functional change. This observation does not preclude an ontogenetic change in prosauropod diets. On the basis of histological evidence, it has been suggested that juvenile *Massospondylus* grew at higher rates than adults (Chinsamy 1993). If this were the case we might expect juveniles to have higher nutritional demands than adults. These demands may have been met by an increased dependence on animal material in juvenile diets.

The teeth of *Yunnanosaurus*

Yunnanosaurus is a large prosauropod dinosaur from the Dark Red Beds of the Lower Jurassic Lower Lufeng Formation of Yunnan (Young 1951). The structure of its teeth is unique among prosauropods (Young 1942; Galton 1985a) in that the tooth crowns are broadly spatulate, labially convex, and generally lack denticles (one or two denticles may be present at the apices of the teeth [Young 1942; pers. obs.]). This unusual pattern, similar to that of sauropods, may indicate that *Yunnanosaurus* relied upon vegetation much more heavily than other prosauropods. Although the teeth of *Yunnanosaurus* are sauropod-like, no other cranial characters support the assignment of this genus to the Sauropoda. Known sauropod skulls possess a suite of autapomorphies (transversely broad snout, contact between the posterior end of the ascending process of the maxilla and the lacrimal, retraction of the external nares, and anteroposterior shortening of the nasals and frontals; Upchurch 1995, 1998; Upchurch and Barrett, this volume), which are absent in *Yunnanosaurus* (Young 1942; pers. obs.).

Simmons (1965) briefly mentioned several isolated teeth from the Lower Lufeng Formation, which he referred to *Yunnanosaurus*. These teeth (referred to hereafter as '*Yunnanosaurus*') have large wear facets along their mesial and distal edges, extremely similar to the wear on some sauropod teeth, which can only have been formed by tooth-to-tooth contact during occlusion (Galton 1985a; Figure 3.6a). These wear facets indicate the presence of precise occlusion in '*Yunnanosaurus*'. The teeth of all other prosauropods lack wear (except in rare instances), indicating that tooth-to-tooth occlusion was absent. The condition seen in '*Yunnanosaurus*' would represent a significant change toward herbivory in the masticatory apparatus of prosauropods (Galton 1985a).

However, referral of these teeth to *Yunnanosaurus* is doubtful (Barrett 1998). The teeth of '*Yunnanosaurus*' have crowns with an almost cylindrical cross-section. The crowns of teeth definitely referable to *Yunnanosaurus* are labiolingually compressed (IVPP V94, V505). Also, the provenance of the

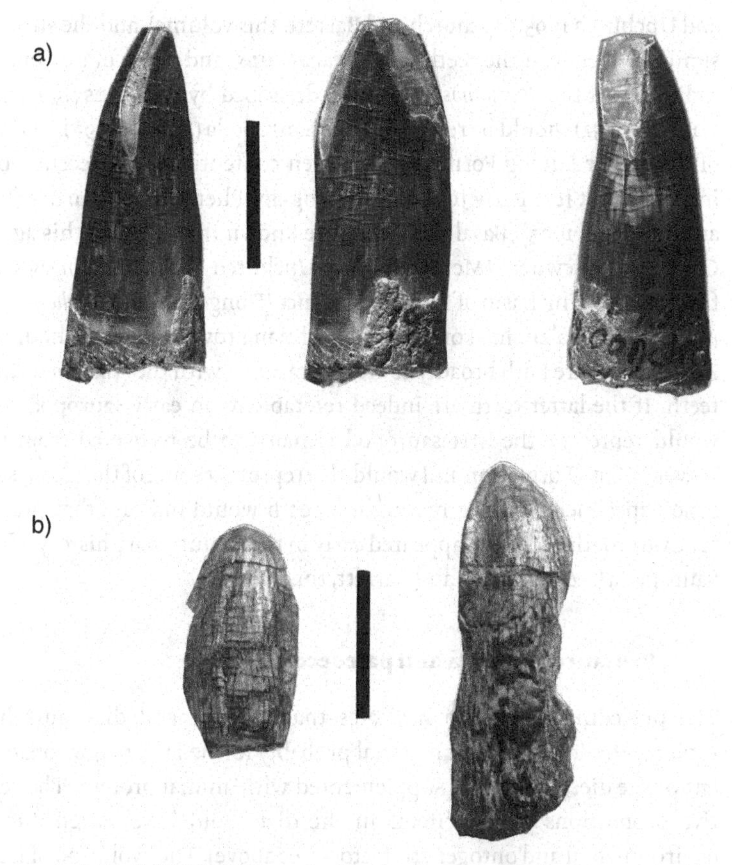

Figure 3.6. a) Tooth (FMNH CUP 2051) referred to *Yunnanosaurus* by Simmons (1965) in mesial, labial, and distal views showing wear facets. b) Posterior teeth of *Yunnanosaurus* (IVPP V94) in labial view. Scale bars each equal 10 mm.

'*Yunnanosaurus*' teeth is unknown in most cases; although it is known that they were collected from the Lower Lufeng Formation, the exact localities and horizons (Dull Purplish or Dark Red Beds) have not been recorded, and no material unquestionably identifiable as *Yunnanosaurus* can be directly associated with these teeth (Simmons 1965). Furthermore, tooth wear has not been described in any of the type specimens of *Yunnanosaurus* spp. (Young 1942, 1951) and was not observed by me in a recent examination of several specimens of *Yunnanosaurus* (IVPP V94, V505; Figure 3.6b).

The general absence of tooth wear in prosauropods, the absence of such wear in several specimens unquestionably referable to *Yunnanosaurus*, the prevalence of this type of wear in sauropods (Calvo 1994; Barrett and Upchurch 1995; Upchurch and Barrett, this volume), and the striking similarity between the teeth of '*Yunnanosaurus*' and those of sauropods, indicate that the '*Yunnanosaurus*' teeth described by Simmons (1965) and Galton (1985a) should be referred to the Sauropoda (Barrett 1998). The age of the Lower Lufeng Formation has been contentious, but recent work indicates that it is Early Jurassic (Hettangian–Pliensbachian) in age (Luo and Wu 1994, 1995). Basal sauropods are known from beds of this age in China and elsewhere (McIntosh 1990; Upchurch 1995). *Kunmingosaurus*, from the Wuding Basin of Yunnan Province (Dong 1992), and *Zizhongosaurus*, from the Da'znzhai Formation of Sichuan Province (Dong, Zhou, and Zhang 1983), are both broadly contemporaneous with the '*Yunnanosaurus*' teeth. If the latter teeth are indeed referable to an early sauropod they would represent the first sauropod remains to be recovered from the Lower Lufeng Formation and would also represent some of the oldest sauropod specimens. Furthermore, these teeth would indicate that precise tooth-to-tooth occlusion appeared early in the evolutionary history of the Sauropoda (see Upchurch and Barrett, this volume).

Prosauropod diets and paleoecology

The preceding discussion indicates that prosauropod dinosaurs had rather catholic diets. Plant material probably formed the major constituent of the diet, which was supplemented with animal protein. The relative proportions of food items in the diet could have varied due to environmental and ontogenetic factors (see above). The evolution of large body size in prosauropods may be correlated with a shift towards an herbivorous diet, as has been suggested for the sauropods (Farlow 1987). Increased body size allows slower passage times through the gut and a longer period of gut fermentation – two processes that may have allowed these animals to eat tough, hard-to-digest vegetation (Farlow 1987). This suggestion is in accordance with our current knowledge of prosauropod paleoenvironments (Table 3.2) where plant material may have been tough and xerophytic. Alternatively, the larger prosauropods (e.g., melanorosaurids, *Plateosaurus*) may have relied more heavily on vegetation than their smaller relatives.

Two distinct prosauropod assemblages were present in the Karoo Basin of southern Africa; one in the Lower Elliot Formation (which is late Carnian or early Norian in age; Olsen and Galton 1984) consisting of *Euskelosaurus*, *Blikanasaurus* and *Melanorosaurus*, and one in the Upper Elliot,

Clarens, and Forest Sandstone formations (which is Hettangian–Sinemurian in age; Olsen and Galton 1984) containing the single genus *Massospondylus* (Haughton 1924; Kitching and Raath 1984). The deposition of these strata took place during a period of increasing aridity. The strata of the Lower Elliot Formation display evidence of some permanent water sources and ephemeral surface water. The Upper Elliot strata consist of playa and braided stream deposits, and the Clarens and Forest Sandstone formations indicate drier conditions with playa and sand dunes (Smith, Eriksson, and Botha 1993). Haughton (1924) and Charig *et al.* (1965) suggested that the change in the prosauropod assemblages was initiated by the changes in climate, though these authors believed that differences in the locomotory capabilities and preferred habitats of the various prosauropod taxa accounted for the differences between the Lower and Upper Elliot assemblages. Here, it is suggested that the relatively mesic environment prevailing during Lower Elliot times may have supported denser vegetational cover than the drier environments present during Upper Elliot/Clarens times and may have offered more opportunities for prosauropod diversification. Unfortunately, few plant fossils are recorded from these formations (Gow *et al.* 1990), and thus vegetational cover must be inferred from climatic evidence.

Many Mid-Jurassic to Early Cretaceous dinosaurian assemblages are characterized by a remarkable diversity of sauropods, and some localities have yielded up to six sauropod genera (Weishampel 1990). This diversity has been attributed to variation in the feeding adaptations of sauropods, including cranial and postcranial characters, which may have allowed niche partitioning between the various sauropod taxa (Barrett and Upchurch 1995; Upchurch and Barrett, this volume). Several prosauropod assemblages also display some degree of taxonomic diversity (Table 3.2). In these cases, it does not seem possible to separate the taxa ecologically as prosauropod feeding mechanisms appear to be remarkably uniform (Galton 1984a, 1985a, 1986). Furthermore, niche separation on the basis of differences in body size does not seem to occur as many sympatric taxa reached similar adult body sizes. However, if prosauropods were facultatively omnivorous, rather than strictly herbivorous, this would have allowed greater dietary variation within and between genera, reducing niche competition and allowing the diversification of similar genera within the same environment.

The supposed generic diversity in at least one prosauropod fauna, that of the Lower Lufeng Formation, may be artificially high. Young (summarized in Young 1951) named six species comprising three genera of prosauropod dinosaurs, *Lufengosaurus*, *Yunnanosaurus*, and *Gyposaurus*. Rozhdestvensky (1965) and Galton (1990) argued that *Lufengosaurus* and *Yunnanosaurus* are valid monotypic genera and that *Gyposaurus* is a subjective junior synonym of *Lufengosaurus*. Most authors have stated that these two genera are distributed throughout the Lower Lufeng Formation in both the Dull Purplish and Dark Red Beds (Young 1951; Weishampel 1990). However, it has been suggested that there are two distinct prosauropod assemblages, the older *Lufengosaurus* fauna (Dull Purplish Beds) and a younger (Dark Red Beds) *Yunnanosaurus* fauna (Dong 1992). The latter possibility seems the most likely; the specimens referable to each genus come from the Dull Purplish Beds (*Lufengosaurus*) and Dark Red Beds (*Yunnanosaurus*), respectively. Furthermore, the stratigraphical ranges of each of these genera were extended into the other unit of the formation by the referral of isolated prosauropod remains to each of these taxa by Young (1951). Much of this referred material is actually not identifiable at lower taxonomic levels; those specimens that do show diagnostic features of either *Lufengosaurus* or *Yunnanosaurus* always come from the horizon from which the holotypes of the type-species of the respective genus was obtained (P. M. Barrett, unpubl. data). The recent description of *Jingshanosaurus* (Zhang and Yang 1994) from the Dull Purplish Beds implies that two prosauropod taxa may have been present during the time of deposition of this unit, but only one (*Yunnanosaurus*) was present in the Dark Red Beds. A taxonomic revision of the prosauropod dinosaurs from the Lower Lufeng Formation is long overdue and is needed to resolve this rather unsatisfactory situation.

Diet in other extinct reptilian groups

Some of the dental features present in prosauropod dinosaurs and iguanine lizards are present in a variety of other dinosaurian taxa, including basal ornithischians and some theropods, and in several non-dinosaurian archosaurs such as crocodyliforms.

Ornithischia

The basal ornithischian *Lesothosaurus* has a differentiated upper dentition comprising 6 premaxillary teeth and 15 or 16 maxillary teeth (Thulborn 1970, 1971; Sereno 1991). The maxillary teeth have low, (in labial or lingual view) triangular crowns that bear coarse denticles along their mesial and

Figure 3.7. Teeth of *Lesothosaurus* in labial view. a) Maxillary teeth (BMNH R8501). b) Premaxillary teeth (BMNH RUB17). Scale bars each equal 1 mm.

distal edges (Figure 3.7a). The tooth crowns become progressively broader in the more posterior maxillary teeth. In contrast, the premaxillary teeth are recurved and lack the large denticles present on the maxillary teeth (Figure 3.7b). The fourth and fifth premaxillary tooth crowns bear very fine serrations along their mesial cutting edges, and the sixth premaxillary tooth bears serrations on both mesial and distal margins

(Thulborn 1970, 1971; Sereno 1991; Figure 3.7b). The similarity of these teeth to those of prosauropods and iguanines (see above), and the absence of many adaptations to herbivory (such as a ventrally displaced jaw joint, precise occlusion, relative movements of the jaws), indicate that basal ornithischians may have been facultatively omnivorous.

Hypsilophodontid and heterodontosaurid ornithopods retain recurved premaxillary teeth (Galton 1974; Crompton and Charig 1962). However, the skulls of hypsilophodontids and other euornithopods possess a pleurokinetic hinge that allowed the production of a transverse power stroke during jaw closure (Norman and Weishampel 1985). This jaw mechanism, and the resulting tooth wear, are most probably associated with an almost exclusively herbivorous diet. Various other features, such as a ventrally displaced jaw joint, also indicate that euornithopods were primarily herbivorous. The premaxillary teeth may have been used for the prehension of plant food or, in rare instances, capture of small prey. Iguanodontians lack premaxillary teeth, have well-developed pleurokinetic hinge systems, and complex dental batteries. These features, and many others, indicate that these dinosaurs were exclusively herbivorous. The diet of heterodontosaurids is more problematic. The jaw action and structure of the cheek teeth (Weishampel 1984; Crompton and Attridge 1986) appear to indicate an herbivorous diet, but the shapes of the premaxillary teeth and of the dentary caniniform tooth suggest that facultative omnivory may have been a possibility. Molnar (1977) proposed that the caniniform tooth was used in intraspecific combat and display, as it is in primitive cervids like *Muntiacus*. This function seems entirely feasible. However, the caniniform teeth of *Heterodontosaurus* are finely serrated, in a manner reminiscent of theropod teeth, whereas the caniniform teeth of cervids lack serrations and possess very sharp carinae (pers. obs.). The presence of serrations on the caniniforms of *Heterodontosaurus* may indicate that they were used during feeding. Interestingly, the cervids with enlarged caniniform teeth are known to take occasional animal prey (see references cited in Jarman 1974). The enigmatic Early Cretaceous ornithischian *Echinodon* has a caniniform anterior maxillary tooth (Galton 1978), but the crown of this tooth is not serrated (BMNH 48209; pers. obs.). This tooth may have performed similar functions to those postulated for the dentary caniniform of *Heterodontosaurus* (see above).

Pachycephalosaurs retained recurved premaxillary teeth and lacked many obvious adaptations to herbivory. The more posterior teeth bear a strong resemblance to those of iguanines, prosauropods, and basal ornithischians. These observations indicate that pachycephalosaurs may also have been facultative omnivores. Stegosaurs and ankylosaurs have simple maxillary and dentary teeth and lack a specialized masticatory apparatus, and some retained slightly recurved premaxillary teeth. On the basis of dental features alone, some of these animals might be regarded as facultative omnivores. However, the large body size, barrel-shaped abdomen, heavily armored bodies, quadrupedal habits, and the lack of a manipulative manus in derived thyreophorans indicate that they had an exclusively herbivorous diet.

Theropoda

Although most theropods possessed a typically 'carnivorous' dentition (recurved, labiolingually compressed tooth crowns with serrated cutting edges), there are several exceptions to this rule. Ornithomimosaurs are either toothless or retain a poorly developed dentition in the anterior portion of the mandible (Barsbold and Osmólska 1990). Oviraptorosaurs are completely toothless (Barsbold, Maryanska, and Osmólska 1990). These groups are usually considered to be carnivorous, either swallowing small prey whole (ornithomimosaurs) or using strong adductor muscles to crush molluscs against the roof of the mouth prior to swallowing (oviraptorosaurs; Barsbold 1983). Cracraft (1971) noted several similarities between the jaw apparatus of caenagnathid oviraptorosaurs and dicynodont therapsids, although he did not explicitly suggest that caenagnathids were herbivorous. Although they accepted the functional similarities between caenagnathids and dicynodonts, Barsbold *et al.* (1990) still considered caenagnathids carnivorous. However, several authors have argued that some of these groups were either herbivorous or omnivorous (Paul 1988; Smith 1992; Sues 1997). As all of these taxa typically lack teeth they will not be further discussed here.

Pelecanimimus, a basal ornithomimosaur from the Early Cretaceous of Spain, possesses over 200 small teeth (Pérez-Moreno *et al.* 1994). The teeth are confined to the anterior portions of the upper and lower jaws. The anterior teeth are recurved, and the posterior teeth have slender, blade-like crowns. Pérez-Moreno *et al.* (1994) argued that this arrangement was intermediate between the dentitions of troodontids, which possess denticulate teeth with moderate interdental spaces (Currie 1987), and the more derived ornithomimids, which lack any teeth. The large number of closely packed tooth crowns in *Pelecanimimus* was interpreted as the functional equivalent of a single cutting edge, similar to that provided by a keratinous beak (Pérez-Moreno *et al.* 1994). If this functional interpretation is accepted, herbivory, omnivory, or carnivory are equally likely for this animal. Extant turtles, which also lack teeth, use the beak for a variety of dietary strategies.

Troodontid theropods are usually thought to be small predators (Osmólska and Barsbold 1990). However, Holtz and Chandler (1994) have argued, on the basis of denticle size and serration density, that *Troodon* may have been omnivorous or even herbivorous. In most theropods, denticle size and serration density fall on the same allometric growth curve. In contrast, troodontid denticles are relatively larger and more widely spaced (Figure 3.2c). This is also the case in basal ornithomimosaurs and therizinosaurs. The structure of the tooth crowns in these groups was found to be more similar to that of 'herbivorous' dinosaurs and iguanines than to that of undoubted carnivores (Holtz and Chandler 1994). The similarity of troodontid and therizinosauroid teeth was also noted by Russell and Dong (1994).

The affinities of therizinosaurs have been the subject of some debate. Some authors have united them with the prosauropods (Sereno 1989), while others have suggested that they were relicts of an alleged prosauropod–ornithischian transition (Paul 1984). However, recent studies suggest that they are derived members of the Theropoda (Russell and Dong 1994; Clark, Perle, and Norell 1994; Sues 1997) as originally suggested by Perle (1979). Paul (1984) argued that therizinosaurs may have been herbivorous based on the inferred presence of 'cheeks,' the retroversion of the pubis (allowing for a larger gut to process plant food), the inferred presence of a keratinous beak, and the similarity of therizinosauroid teeth to those of prosauropods and basal ornithischians. The coarsely serrated, mesiodistally expanded tooth crowns indicate that omnivory is also a possibility.

Non-dinosaurian archosaurs

Most crocodilians have simple conical teeth, with tooth crowns that are not expanded relative to the width of the root and taper to a point. There is typically little variation in tooth form along the tooth row (Romer 1956). By analogy with extant crocodilians, extinct forms have been interpreted as having an exclusively carnivorous diet. However, *Chimaerasuchus paradoxus*, a recently described crocodyliform from the Early Cretaceous of China, has molariform maxillary teeth that are strikingly similar to those of tritylodontid synapsids (Wu, Sues, and Sun 1995). The molariform tooth crowns possess several rows of low, curved cusps and are mesiodistally and labiolingually expanded. *Chimaerasuchus* also has procumbent premaxillary teeth. This combination of dental characters led the authors to conclude that 'the structure of the maxillary molariforms is consistent with a specialised diet including fibrous material and may indicate at least facultative herbivory in *Chimaerasuchus*. The procumbent premaxillary . . . teeth were presumably involved in seizing food items' (Wu *et al.* 1995). This combination of dental features seems ideally suited to omnivory. The mandible of *Chimaerasuchus* was capable of propalinal movements, which may have been involved in processing food (Wu *et al.* 1995). However, the lack of wear on the teeth indicates that food processing would not have been intensive. Propalinal jaw actions are generally found in herbivores and are associated with the grinding or shearing of plant material (King 1996). However, this does not preclude the development of propaliny in an omnivore, which would take at least some vegetation in its diet.

Another unusual crocodilyform, *Malawisuchus* from the Early Cretaceous of Malawi, also has an unusual, mammal-like dentition and appears to have been capable of propalinal jaw movements (Clark, Jacobs, and Downs 1989; Gomani 1997). The anterior teeth of this form are conical and recurved whereas the posterior maxillary and dentary teeth each have a single large central cusp and a cuspidate cingulum. It has been proposed that the molariform posterior maxillary and dentary teeth were employed in the processing of prey items (using proal motion of the mandible to produce shearing between the upper and lower dentitions) and that the anterior teeth ('caniniforms') were used for prey capture (Clark *et al.* 1989; Gomani 1997). King (1996) argued that *Malawisuchus* may have been herbivorous due to the possession of molariform teeth and the ability to move the jaw fore-and-aft. However, although the teeth of *Malawisuchus* are superficially mammal-like, they bear a stronger resemblance to the carnassial teeth of carnivorous mammals than to the grinding teeth of herbivores. Furthermore, although proal jaw actions are more common in herbivores, there are some instances of propaliny in animals such as *Sphenodon* (Robinson 1976), which have a diet of insects and other small prey. Therefore, fore-and-aft jaw motion does not preclude a carnivorous diet. Indeed, in some cases it may be advantageous to allow some translational movements of the mandible at the jaw joint in carnivorous animals in order to get a firmer grip on prey (by using a palinal movement to dig the teeth further into the prey), or to allow shearing between the upper and lower dentitions (to cut through flesh and bone). It seems likely that *Malawisuchus* was a specialized predator, as originally inferred by Clark *et al.* (1989), rather than a herbivore.

The fossilized gut contents of the Late Permian archosauromorph reptile *Protorosaurus* provide a further indication that tooth form does not always reflect diet (Munk and Sues 1993). The teeth of *Protorosaurus* are recurved and conical (i.e., 'carnivorous'), but the abdominal cavity in two specimens contains an *in situ* gastric mill and many ovules of the conifer *Ullmannia* and pteridosperms. Although the ovules may have been ingested accidentally, the combined presence of the ovules and a gastric mill support the suggestion that *Protorosaurus* was an omnivore (Munk and Sues 1993).

Evolution of dinosaurian herbivory

All postulated dinosaurian outgroups (*Pseudolagosuchus*, *Marasuchus*, *Lagerpeton*, Pterosauria, *Scleromochlus*, Crurotarsi; Bennett 1996; Gower and Wilkinson 1996; Novas 1996) are carnivorous. This leads logically to the assumption that the 'ancestral' dinosaur was a carnivore (implying retention of the primitive dinosauriform condition). Carnivory was supposedly retained in the Theropoda, whereas herbivory appeared twice in the evolutionary history of the Dinosauria – once at the base of Ornithischia and once at the base of Sauropodomorpha (e.g., Sereno 1997). This view is supported by the observation that both *Herrerasaurus* and *Staurikosaurus* possess a classical carnivore dentition and other cranial features (e.g., intramandibular joint, Sereno and Novas 1994) indicative of carnivory. These taxa were regarded as 'basal dinosaurs' that represented successive outgroups to a monophyletic clade of Saurischia plus Ornithischia (Brinkman and Sues 1987; Benton 1990; Novas 1992). Precladistic views of dinosaurian evolution present a similar picture of dietary evolution, with ornithischians, sauropodomorphs and theropods independently derived from different carnivorous 'thecodontian' taxa (e.g., Charig *et al.* 1965; Charig 1979).

In an attempt to test the assumption that carnivory represents the ancestral dinosaurian condition, inferred dinosaurian diets (carnivory, omnivory, herbivory) were mapped on to a cladogram of dinosaurian interrelationships using MacClade 3.01 (Maddison and Maddison 1992; Figure 3.8). The cladogram used for this purpose is based on the analysis

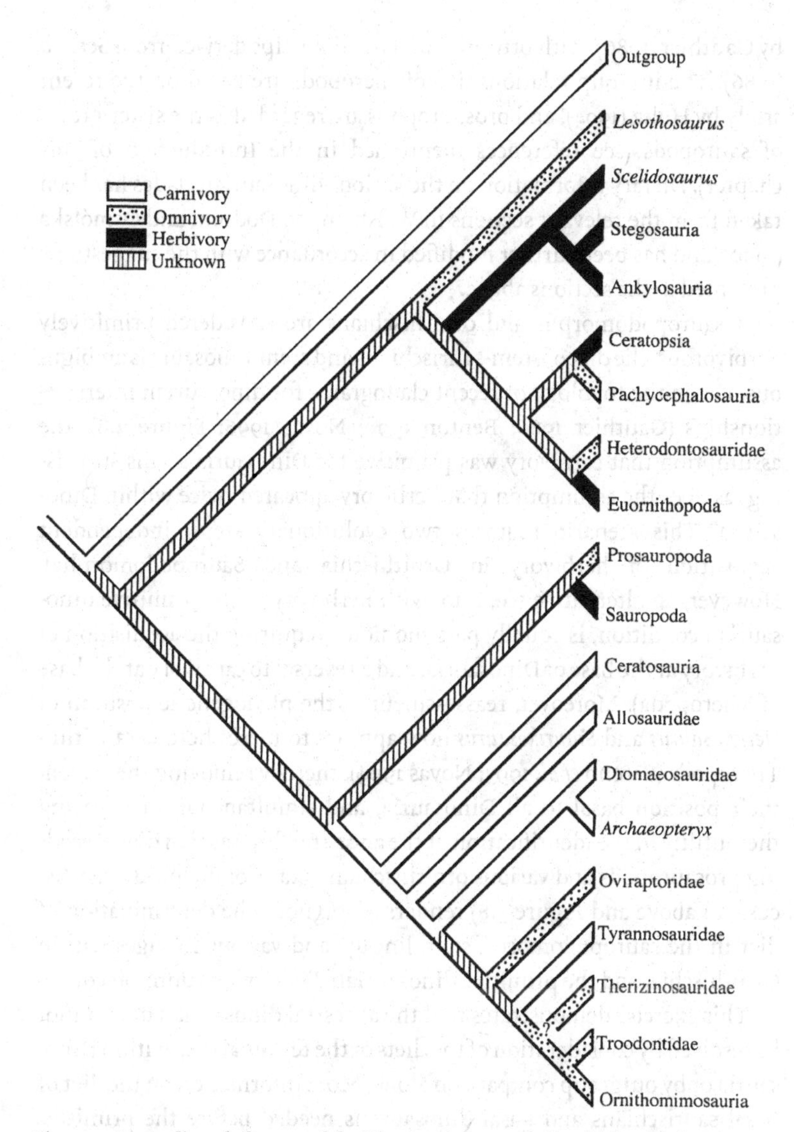

Figure 3.8. Outline phylogeny of the Dinosauria (sources: Gauthier 1986; Sereno 1986, 1989; Holtz 1994) with distribution of inferred dietary habits mapped on the cladogram (see text for discussion).

by Gauthier (1986), with ornithischian relationships derived from Sereno (1986). The ingroup relationships of theropods are based on the recent study by Holtz (1994), and prosauropods are regarded as the sister-group of sauropods (see references mentioned in the Introduction of this chapter). Dietary information on the various dinosaurian clades has been taken from the relevant sections in Weishampel, Dodson, and Osmólska (1990), and has been further modified in accordance with the suggestions presented in the sections above.

If sauropodomorphs and ornithischians are considered primitively herbivorous, the diet of stem-saurischians and stem-dinosaurs is ambiguous. Given the topology of recent cladograms for dinosaurian interrelationships (Gauthier 1986; Benton 1990; Novas 1996; Figure 3.8), the assumption that carnivory was primitive for Dinosauria seems surprising, as does the assumption that herbivory appeared twice within Dinosauria. This scenario requires two evolutionary steps (independent acquisition of herbivory in Ornithischia and Sauropodomorpha). However, an alternative scenario, with herbivory as the primitive dinosaurian condition, is equally parsimonious (requiring the acquisition of herbivory at the base of Dinosauria, and a reversal to carnivory at the base of Theropoda). Moreover, reassessment of the phylogenetic position of *Herrerasaurus* and *Staurikosaurus* now appears to place these taxa within Theropoda (Sereno *et al.* 1993; Novas 1994), thereby removing them from their position basal to all Dinosauria, and simultaneously decreasing their utility in the identification of the ancestral dinosaurian diet. Recoding prosauropods and various ornithischian taxa as omnivorous (see discussion above and Figure 3.8) renders ambiguous the determination of diet in the sauropodomorph stem-lineage and various lineages within Ornithischia, and the primitive dinosaurian diet remains unresolved.

This exercise demonstrates that the ancestral dinosaurian diet cannot be resolved by consideration of the diets of the terminal taxa within Dinosauria or by outgroup comparison alone. More information on the diet of basal saurischians and basal dinosaurs is needed before the primitive dinosaurian diet can be confidently inferred. In this case, parsimony cannot distinguish between the competing hypotheses of primitive dinosaurian omnivory, herbivory, or carnivory, suggesting that it may not be wise to rely solely on parsimony-based arguments when attempting to reconstruct the evolution of biological systems, especially if all of the taxa under consideration are now extinct. Other phylogenetic approaches, such as the Extant Phylogenetic Bracket method (Witmer 1995; Weishampel 1995), are useful in deducing features of extinct organisms, but the latter approach relies on the use of extant organisms as outgroups, a luxury that is denied to groups where most of the ingroup members, as well as the immediate outgroups, are extinct.

Although parsimony indicates that, primitively, herbivory, carnivory or omnivory were equally probable, other factors indicate that the hypothesis of primitive herbivory can be rejected. Obligate herbivory requires a substantial number of cranial, dental, and physiological modifications (see Sues and Reisz [1998] for a recent review), and it is considered biologically improbable that a committed herbivore would revert to carnivory. By contrast, the suggestion that carnivory or omnivory was the ancestral dinosaurian condition is supported by the presence of 'carnivorous' features (e.g., recurved, serrated premaxillary teeth) and the lack of derived 'herbivorous' features (e.g., a ventrally displaced jaw joint) in the basal members of all major dinosaurian clades. This observation adds support to the suggestion that many of the adaptations commonly regarded as characteristic of herbivorous feeding complexes may have originated in the presence of continued carnivorous or omnivorous habits.

The distribution of diets on the cladogram (Figure 3.8) indicates that obligate herbivory may have had several independent origins within Dinosauria, rather than two as usually supposed. For example, herbivory may have evolved at least three times within Ornithischia. Similarly, omnivory may either have had multiple origins within Dinosauria or may have been primitive for the group as a whole. In either case, omnivory appears to be more widespread than previously acknowledged.

Conclusions

Perhaps tooth form, at least in the case of reptilian herbivores, is not the reliable predictor of diet that it was once thought to be. The examples listed above reflect the difficulty of relating tooth form to diet in many taxa. Omnivory is not recognized, or even suggested, in many extinct taxa due to the tendency of most authors to place animals at either end of the carnivory/herbivory spectrum. This is due, in part, to the application of an unsatisfactory paradigm for herbivory (iguanine tooth structure). Another reason for the reluctance of authors to infer omnivory is that dental correlates specific to omnivory have never been identified. Perhaps the combination of recurved anterior teeth and more 'herbivore-like' posterior teeth could represent such a correlate, at least in the major taxa (Dinosauria, Lepidosauria, Crocodylomorpha) mentioned above. The anterior teeth would serve in the procurement and dismemberment of prey items or carrion, or the procurement of plant material, whereas the posterior teeth would have been suited to the shearing of both plant and animal fodder. Members of these taxa that are closer to the herbivorous end of the dietary spectrum may be expected to lose their premaxillary teeth, or to modify them to aid in the prehension or shearing of plant material. Other features of the dentition in these taxa may display more extensive modifications for herbivory, such as the possession of dental batteries, the presence of extensive tooth wear (associated with modifications of jaw motion), and the development of molariform teeth.

Some authors have argued that it may be unrealistic to compare prosauropods, or for that matter other taxa, with iguanine lizards, due to differences in body size and remote phylogenetic relationship (Gow *et al.* 1990). However, the absence of other extant analogues almost forces us to make such comparisons. This approach does at least allow some empirically based paleobiological inferences to be made. Also, the detailed similarities between the dentitions of iguanine lizards and the other taxa discussed above suggest that these comparisons can be meaningful.

If many of the 'reptilian' taxa currently thought to be carnivorous or herbivorous could be identified as omnivores, this would have important implications for those working on the trophic dynamics of late Paleozoic and Mesozoic terrestrial ecosystems.

Acknowledgments

This manuscript benefited from many discussions with Drs P. Upchurch and D. B. Norman. I would also like to thank the following for discussion on the various issues raised during the preparation of this chapter: Drs A. B. Arcucci, E. N. Arnold, S. E. Evans, D. J. Gower, Z. Luo, H.-D. Sues, M. A. Taylor and D. M. Unwin, and S. Finney and J. Pearson. Drs D. B. Weishampel, R. R. Reisz, P. M. Galton, D. B. Norman, P. Upchurch, A. E. Friday, H.-D. Sues and M. J. Benton provided helpful reviews of the manuscript. Dr A. C. Milner and S. Chapman (Natural History Museum, London), R. Symonds (University Museum of Zoology, Cambridge), Professor Dong Zhiming (Institute of Vertebrate Paleontology and Paleoanthropology, Beijing), W. Simpson (Field Museum of Natural History, Chicago), C. R. Schaff (Museum of Comparative Zoology, Harvard) and Drs B. Rubidge and C. Gow (Bernard Price Institute for Palaeontological Research, Johannesburg) granted access to specimens in their care. Many thanks to the Lethaia Foundation and Dr P. M. Galton for permission to reproduce the illustrations used in Figure 3.1b,c, and to Dr P. J. Currie and the Society of Vertebrate Paleontology for permission to use the drawing in Figure 3.2c. Figure 3.8 was created by Dr P. Upchurch. Sincere thanks to Dudley Simons for photographic assistance. The Dinosaur Society and Cambridge Philosophical Society provided funds for visits to the People's Republic of China, and Trinity College (Cambridge) supported a research trip to the United States. This work was initiated while in receipt of a NERC studentship (GT4/93/128/G) and continued as a Research Fellow of Trinity College, Cambridge.

References

Barrett, P. M. (1998). *Herbivory in the Non-Avian Dinosauria*. Unpublished Ph.D. dissertation, University of Cambridge.

Barrett, P. M., and Upchurch, P. (1995). Sauropod feeding mechanisms: their bearing on palaeoecology. In *6th Symposium on Mesozoic Terrestrial Ecosystems and Biotas, Short Papers*, ed. A. Sun, and Y. Wang, pp. 107–110. Beijing: China Ocean Press.

Barsbold, R. (1983). [Carnivorous dinosaurs from the Cretaceous of Mongolia]. *Sovm. Sov.-Mong. Paleontol. Eksped. Trudy* 19:1–117 (in Russian with English abstract).

Barsbold, R., Maryanska, T., and Osmólska, H. (1990). Oviraptorosauria. In *The Dinosauria*, ed. D. B. Weishampel, P. Dodson, and H. Osmólska, pp. 249–258. Berkeley: University of California Press.

Barsbold, R., and Osmólska, H. (1990). Ornithomimosauria. In *The Dinosauria*, ed. D. B. Weishampel, P. Dodson, and H. Osmólska, pp. 225–244. Berkeley: University of California Press.

Bennett, S. C. (1996). The phylogenetic position of the Pterosauria within the Archosauromorpha. *Zool. J. Linn. Soc. Lond.* 118:261–309.

Benton, M. J. (1990). Origin and interrelationships of dinosaurs. In *The Dinosauria*, ed. D. B. Weishampel, P. Dodson, and H. Osmólska, pp. 11–30. Berkeley: University of California Press.

Bonaparte, J. F. (1972). Los tetrapodos del sector superior de la Formacion Los Colorados, La Rioja, Argentina (Triásico Superior). *Opera Lilloana* 22:1–183.

Brinkman, D. B., and Sues, H.-D. (1987). A staurikosaurid dinosaur from the Upper Triassic Ischigualasto Formation of Argentina and the relationships of the Staurikosauridae. *Palaeontology* 30:493–503.

Calvo, J. O. (1994). Jaw mechanics in sauropod dinosaurs. *GAIA* 10:183–194.

Charig, A. (1979). *A New Look at the Dinosaurs*. London: British Museum (Natural History).

Charig, A. J., Attridge, J., and Crompton, A. W. (1965). On the origin of the sauropods and the classification of the Saurischia. *Proc. Linn. Soc. Lond.* 176:197–221.

Chinsamy, A. (1993). Bone histology and growth trajectory of the prosauropod dinosaur *Massospondylus carinatus* Owen. *Modern Geol.* 18:319–329.

Clark, J. M., and Fastovsky, D. E. (1986). Vertebrate biostratigraphy of the Glen Canyon

Group in northern Arizona. In *The Beginning of the Age of Dinosaurs*, ed. K. Padian, pp. 285–301. Cambridge and New York: Cambridge University Press.

Clark, J. M., Jacobs, L. L., and Downs, W. R. (1989). Mammal-like dentition in a Mesozoic crocodylian. *Science* 244:1064–1066.

Clark, J. M., Perle, A., and Norell, M. A. (1994). The skull of *Erlikosaurus andrewsi*, a Late Cretaceous 'segnosaur' (Theropoda: Therizinosauridae) from Mongolia. *Am. Mus. Novitates* 3115:1–39.

Colbert, E. H. (1962). *Dinosaurs. Their Discovery and Their World*. New York: E. P. Dutton & Co., Inc.

Colbert, E. H. (1981). A primitive ornithischian dinosaur from the Kayenta Formation of Arizona. *Bull. Mus. N. Arizona* 53:1–61.

Cooper, M. R. (1981). The prosauropod dinosaur *Massospondylus carinatus* Owen from Zimbabwe: its biology, mode of life and phylogenetic significance. *Occas. Pap. Natl. Mus. Rhodesia, B, Nat. Sci.* 6:689–840.

Cracraft, J. (1971). Caenagnathiformes: Cretaceous birds convergent in jaw mechanism to dicynodont reptiles. *J. Paleont.* 45:805–809.

Crompton, A. W., and Attridge, J. (1986). Masticatory apparatus of the larger herbivores during Late Triassic and Early Jurassic times. In *The Beginning of the Age of Dinosaurs*, ed. K. Padian, pp. 223–236. Cambridge and New York: Cambridge University Press.

Crompton, A. W., and Charig, A. J. (1962). A new ornithischian from the Upper Triassic of South Africa. *Nature* 196:1074–1077.

Cuny, G., and Ramboer, G. (1991). Nouvelles donnees sur la faune et l'age de Saint-Nicolas-de-Port. *Rev. Paléobiol.* 10:69–78.

Currie, P. J. (1987). Bird-like characteristics of the jaws and teeth of troodontid theropods (Dinosauria, Saurischia). *J. Vert. Paleont.* 7:72–81.

Dong, Z. (1992). *The Dinosaurian Faunas of China*. Berlin: Springer-Verlag.

Dong, Z., Zhou, S., and Zhang, Y. (1983). [Dinosaur remains of Sichuan Basin.] *Palaeontol. Sin., n. s., C* 23:1–145 (in Chinese with English abstract).

Farlow, J. O. (1987). Speculations about the diet and digestive physiology of herbivorous dinosaurs. *Paleobiology* 13:60–72.

Frey, F. L. (1986). Feeding and nutritional diseases of reptiles. In *Zoo and Wild Animal Medicine*, 2nd Edition, ed. M. E. Fowler, pp. 139–151. Philadelphia: W. B. Saunders and Company.

Furness, R. W. (1988a). Predation on ground nesting seabirds by populations of red deer *Cervus elephas* and sheep *Ovis*. *J. Zool.* 216:565–573.

Furness, R. W. (1988b). The predation of tern chicks by sheep. *Bird Study* 35:199–202.

Furness, R. W. (1989). Not by grass alone. *Natural History* 1989(12):8–12.

Galton, P. M. (1973). The cheeks of ornithischian dinosaurs. *Lethaia* 6:67–89.

Galton, P. M. (1974). The ornithischian dinosaur *Hypsilophodon* from the Wealden of the Isle of Wight. *Bull. Brit. Mus. (Nat. Hist.), Geol.* 25:1–152.

Galton, P. M. (1976). Prosauropod dinosaurs of North America. *Postilla* 169:1–98.

Galton, P. M. (1978). Fabrosauridae, the basal family of ornithischian dinosaurs (Reptilia: Ornithopoda). *Paläont. Z.* 52:138–159.

Galton, P. M. (1984a). Cranial anatomy of the prosauropod dinosaur *Plateosaurus* from the Knollenmergel (Middle Keuper, Upper Triassic) of Germany. I. Two complete skulls from Trossingen/Württ. with notes on the diet. *Geol. Palaeontol.* 18:139–171.

Galton, P. M. (1984b). An early prosauropod dinosaur from the Upper Triassic of Nordwürttemberg, West Germany. *Stuttgarter Beitr. Naturk., Ser. B* 106:1–25.

Galton, P. M. (1985a). Diet of prosauropods from the Late Triassic and Early Jurassic. *Lethaia* 18:105–123.

Galton, P. M. (1985b). The poposaurid thecodontian *Teratosaurus suevicus* V. MEYER, plus referred specimens mostly based on prosauropod dinosaurs, from the Middle Stubensandstein (Upper Triassic) of Nordwürttemberg. *Stuttgarter Beitr. Naturk., Ser. B* 116:1–29.

Galton, P. M. (1986). Herbivorous adaptations of Late Triassic and Early Jurassic dinosaurs. In *The Beginning of the Age of Dinosaurs*, ed. K. Padian, pp. 203–221. Cambridge and New York: Cambridge University Press.

Galton, P. M. (1990). Basal Sauropodomorpha–Prosauropoda. In *The Dinosauria*, ed. D. B. Weishampel, P. Dodson, and H. Osmólska, pp. 320–344. Berkeley: University of California Press.

Gauffre, F.-X. (1995). Phylogeny of prosauropod dinosaurs. *J. Vert. Paleont.* 15(suppl. to 3):31A.

Gauthier, J. A. (1986). Saurischian monophyly and the origin of birds. *Mem. Calif. Acad. Sci.* 8:1–55.

Gomani, E. M. (1997). A crocodyliform from the Early Cretaceous Dinosaur Beds, northern Malawi. *J. Vert. Paleont.* 17:280–294.

Gow, C. E., Kitching, J. W., and Raath, M. A. (1990). Skulls of the prosauropod dinosaur *Massospondylus carinatus* Owen in the collections of the Bernard Price Institute for Palaeontological Research. *Palaeontol. Afr.* 27:45–58.

Gower, D. J., and Wilkinson, M. (1996). Is there any consensus on basal archosaur phylogeny? *Proc. R. Soc. Lond.* B 263:1399–1406.

Harshbarger, J. W., Repenning, C. A., and Irwin, J. H. (1957). Stratigraphy of the uppermost Triassic and the Jurassic rocks of the Navajo County. *Prof. Pap. U.S. Geol. Surv.* 291:1–74.

Haughton, S. H. (1924). The fauna and stratigraphy of the Stormberg Series. *Ann. S. Afr. Mus.* 12:323–497.

Holtz, T. H., Jr. (1994). The phylogenetic position of the Tyrannosauridae: implications for theropod systematics. *J. Paleont.* 68:1100–1117.

Holtz, T. H., Jr., and Chandler, C. L. (1994). Denticle morphometrics and a possibly different life habit for the theropod dinosaur *Troodon*. *J. Vert. Paleont.* 14 (suppl. to 3):30A.

Hotton, N. III. (1955). A survey of the adaptive relationships of dentition to diet in the North American Iguanidae. *Am. Mid. Nat.* 53:88–114.

Huene, F. von. (1932). Die fossile Reptil-Ordnung Saurischia, ihre Entwicklung und Geschichte. *Monogr. Geol. Paläontol.* 1, 4:1–361.

Jarman, P. J. (1974). The social organisation of antelope in relation to their ecology. *Behaviour* 58:215–267.

Jenkins, F. A., Jr., Shubin, N. H., Amaral, W. W., Gatesy, S. M., Schaff, C. R., Clemmensen, L. B., Downs, W. R., Davidson, A. R., Bonde, N., and Osbæck, F. (1994). Late Triassic continental vertebrates and depositional environments of the Fleming Fjord Formation, Jameson Land, East Greenland. *Medd. Grønland, Geosci.* 32:1–25.

Kermack, D. (1984). New prosauropod material from South Wales. *Zool. J. Linn. Soc. Lond.* 82:101–117.

King, G. M. (1996). *Reptiles and Herbivory*. London: Chapman and Hall.

Kitching, J. W., and Raath, M. A. (1984). Fossils from the Elliot and Clarens formations (Karoo Sequence) of the north-eastern Cape, Orange Free State and Lesotho, and a suggested biozonation based on tetrapods. *Palaeontol. Afr.* 25:111–125.

Lauder, G. V. (1995). On the inference of function from structure. In *Functional Morphology in Vertebrate Paleontology*, ed. J. J. Thomason, pp. 1–18. Cambridge and New York: Cambridge University Press.

Lull, R. S. (1953). Triassic life of the Connecticut Valley. Revised. *Bull. Conn. State Geol. Nat. Hist. Surv.* 81:1–331.

Luo, Z., and Wu, X-c. (1994). The small tetrapods of the Lower Lufeng Formation, Yunnan, China. In *In the Shadow of the Dinosaurs*, ed. N. C. Fraser and H.-D. Sues, pp. 251–270. Cambridge and New York: Cambridge University Press.

Luo, Z., and Wu, X-c. (1995). Correlation of vertebrate assemblages of the Lower Lufeng Formation, Yunnan, China. In *Sixth Symposium on Mesozoic Terrestrial Ecosystems and Biotas, Short Papers*, ed. A. Sun and Y. Wang, pp. 83–88. Beijing: China Ocean Press.

Maddison, W. P., and Maddison, D. R. (1992). *MacClade, version 3.01*. Sunderland, MA: Sinauer.

Mantell, G. A. (1825). Notice on the *Iguanodon*, a newly discovered fossil reptile, from the sandstone of Tilgate Forest, in Sussex. *Phil. Trans. R. Soc. Lond.* 115:176–186.

McIntosh, J. S. (1990). Sauropoda. In *The Dinosauria*, ed. D. B. Weishampel, P. Dodson, and H. Osmólska, pp. 345–401. Berkeley: University of California Press.

Molnar, R. E. (1977). Analogies in the evolution of combat and display structures in ornithopods and ungulates. *Evol. Theory* 3:165–190.

Montanucci, R. R. (1968). Comparative dentition in four iguanid lizards. *Herpetologica* 24:305–315.

Munk, W., and Sues, H.-D. (1993). Gut contents of *Parasaurus* (Pareiasauria) and *Protorosaurus* (Archosauromorpha) from the Kupferschiefer (Upper Permian) of Hessen, Germany. *Paläontol. Z.* 67:169–176.

Norman, D. B., and Weishampel, D. B. (1985). Ornithopod feeding mechanisms: their bearing on the evolution of herbivory. *Am. Nat.* 126:151–164.

Novas, F. E. (1992). Phylogenetic relationships of the basal dinosaurs, the Herrerasauridae. *Palaeontology* 35:51–62.

Novas, F. E. (1994). New information on the systematics and postcranial skeleton of *Herrerasaurus ischigualastensis* (Theropoda: Herrerasauridae) from the Ischigualasto Formation (Upper Triassic) of Argentina. *J. Vert. Paleont.* 13:400–423.

Novas, F. E. (1996). Dinosaur monophyly. *J. Vert. Paleont.* 16:723–741.

Olsen, P. E., and Galton, P. M. (1984). A review of the reptile and amphibian assemblages from the Stormberg of southern Africa, with special emphasis on the footprints and the age of the Stormberg. *Palaeontol. Afr.* 25:87–110.

Osmólska, H., and Barsbold, R. (1990). Troodontidae. In *The Dinosauria*, ed. D. B. Weishampel, P. Dodson, and H. Osmólska, pp. 259–268. Berkeley: University of California Press.

Paul, G. S. (1984). The segnosaurian dinosaurs: relics of the prosauropod–ornithischian transition? *J. Vert. Paleont.* 4:507–515.

Paul, G. S. (1988). *Predatory Dinosaurs of the World: A Complete Illustrated Guide*. New York: Simon and Schuster.

Pérez-Moreno, B. P., Sanz, J. L., Buscalioni, A. D., Moratalla, J. J., Ortega, F., and Rasskin-Gutman, D. (1994). A unique multi-toothed ornithomimosaur dinosaur from the Lower Cretaceous of Spain. *Nature* 370:363–367.

Perle, A. (1979). [Segnosauridae – a new family of theropods from the Late Cretaceous of Mongolia.] *Sovm. Sov.-Mong. Paleontol. Eksped. Trudy* 8:45–55 (in Russian).

Purcell, S. W., and Bellwood, D. R. (1993). A functional analysis of food procurement in two surgeonfish species, *Acanthurus nigrofuscus* and *Ctenochaetus striatus* (Acanthuridae). *Environ. Biol. Fishes* 37:139–159.

Queiroz, K. de (1987). Phylogenetic systematics of iguanine lizards. *Univ. Calif. Publ. Zool.* 118:1–203.

Robinson, P. L. (1957). The Mesozoic fissures of the Bristol Channel area and their vertebrate faunas. *Zool. J. Linn. Soc. Lond.* 43:260–283.

Robinson, P. L. (1976). How *Sphenodon* and *Uromastyx* grow their teeth and use them. In *Morphology and Biology of Reptiles*, ed. A. d'A. Bellairs and C. B. Cox, pp. 43–64. London: Academic Press.

Romer, A. S. (1956). *Osteology of the Reptiles*. Chicago: University of Chicago Press.

Romer, A. S. (1966). *Vertebrate Paleontology*. 3rd Edition. Chicago: University of Chicago Press.

Rozhdestvensky, A. K. (1965). [Growth changes and some problems of systematics of Asian dinosaurs.] *Paleontol. Zh.* 1965:95–109 (in Russian).

Russell, D. A., and Dong, Z. (1994). The affinities of a new theropod from the Alxa Desert, Inner Mongolia, People's Republic of China. *Can. J. Earth Sci.* 30:2107–2127.

Sander, P. M. (1992). The Norian *Plateosaurus* bonebeds of central Europe and their taphonomy. *Palaeogeogr. Palaeoclimatol. Palaeoecol.* 93:255–299.

Sereno, P. C. (1986). Phylogeny of the bird-hipped dinosaurs (Order Ornithischia). *Nat. Geog. Res.* 2:234–256.

Sereno, P. C. (1989). Prosauropod monophyly and basal sauropodomorph phylogeny. *J. Vert. Paleont.* 9(suppl. to 3):38A.

Sereno, P. C. (1991). *Lesothosaurus*, 'fabrosaurids', and the early evolution of Ornithischia. *J. Vert. Paleont.* 11:168–197.

Sereno, P. C. (1997). The origin and evolution of dinosaurs. *Ann. Rev. Earth Planet. Sci.* 25:435–489.

Sereno, P. C., Forster, C. A., Rogers, R. R., and Monetta, A. M. (1993.) Primitive dinosaur skeleton from Argentina and the early evolution of Dinosauria. *Nature* 361:64–66.

Sereno, P. C. and Novas, F. E. (1994). The skull and neck of the basal theropod *Herrerasaurus ischigualastensis*. *J. Vert. Paleont.* 13:451–476.

Shubin, N. H., Olsen, P. E., and Sues, H.-D. (1994). Early Jurassic small tetrapods from the McCoy Brook Formation of Nova Scotia, Canada. In *In the Shadow of the Dinosaurs*, ed. N. C. Fraser and H.-D. Sues, pp.242–250. Cambridge and New York: Cambridge University Press.

Simmons, D. J. (1965). The non-therapsid reptiles of the Lufeng Basin, Yunnan, China. *Fieldiana, Geol.* 15:1–93.

Smith, D. (1992). The type specimen of *Oviraptor philoceratops*, a theropod dinosaur from the Upper Cretaceous of Mongolia. *N. Jb. Geol. Paläontol., Abh.* 186:365–388.

Smith, R. M. H., Eriksson, P. G., and Botha, W. J. (1993). A review of the stratigraphy and sedimentary environments of the Karoo-aged basins of southern Africa. *J. Afr. Earth Sci.* 16:143–169.

Smoot, J. P. (1991). Sedimentary facies and depositional environments of early Mesozoic Newark Supergroup basins, eastern North America. *Palaeogeogr. Palaeoclimatol. Palaeoecol.* 84:369–423.

Sues, H.-D. (1997). On *Chirostenotes*, a Late Cretaceous oviraptorosaur (Dinosauria: Theropoda) from western North America. *J. Vert. Paleont.* 17:698–716.

Sues, H.-D., and Reisz, R. R. (1998). Origins and early evolution of herbivory in tetrapods. *Trends Ecol. Evol.* 13:141–145.

Swinton, W. E. (1934). *The Dinosaurs. A Short History of a Great Group of Extinct Reptiles.* London: Thomas Murby and Company.

Throckmorton, G. S. (1976). Oral food processing in two herbivorous lizards, *Iguana iguana* (Iguanidae) and *Uromastix aegyptius* (Agamidae). *J. Morphol.* 148:363–390.

Thulborn, R. A. (1970). The skull of *Fabrosaurus australis*, a Triassic ornithischian dinosaur. *Palaeontology* 13:414–432.

Thulborn, R. A. (1971). Tooth wear and jaw action in the Triassic ornithischian dinosaur *Fabrosaurus*. *J. Zool., Lond.* 164:165–179.

Upchurch, P. (1993). *The Anatomy, Phylogeny, and Systematics of the Sauropod Dinosaurs.* Unpublished Ph.D. Dissertation, University of Cambridge.

Upchurch, P. (1995). The evolutionary history of the sauropod dinosaurs. *Phil. Trans. R. Soc. Lond., B* 349:365–390.

Upchurch, P. (1998). The phylogenetic relationships of sauropod dinosaurs. *Zool. J. Linn. Soc. Lond.* 124:43–103.

Wallach, J. D., and Boever, W. J. (1983). *Diseases of Exotic Animals.* Philadelphia: W. B. Saunders and Company.

Weishampel, D. B. (1984). Evolution of jaw mechanisms in ornithopod dinosaurs. *Adv. Anat. Embryol. Cell Biol.* 87:1–110.

Weishampel, D. B. (1990). Dinosaurian distribution. In *The Dinosauria*, ed. D. B. Weishampel, P. Dodson, and H. Osmólska, pp. 63–139. Berkeley: University of California Press.

Weishampel, D. B. (1995). Fossils, function, and phylogeny. In *Functional Morphology in Vertebrate Paleontology*, ed. J. J. Thomason, pp. 34–54. Cambridge and New York: Cambridge University Press.

Weishampel, D. B., Dodson, P., and Osmólska, H. (eds.) (1990). *The Dinosauria.* Berkeley: University of California Press.

Winkler, D. A., Jacobs, L. L., Congleton, J. D., and Downs, W. R. (1991). Life in a sand sea: biota from Jurassic interdunes. *Geology* 19:889–892.

Witmer, L. M. (1995). The extant phylogenetic bracket and the importance of reconstructing soft tissues in fossils. In *Functional Morphology in Vertebrate Paleontology*, ed. J. J. Thomason, pp. 19–33. Cambridge and New York: Cambridge University Press.

Wu, X.-c., Sues, H.-D., and Sun, A. (1995). A plant-eating crocodyliform reptile from the Lower Cretaceous of China. *Nature* 376:678–680.

Young, C. C.(1942). *Yunnanosaurus huangi* Young (gen. et sp. nov.), a new Prosauropoda from the Red Beds at Lufeng, Yunnan. *Bull. Geol. Soc. China* 22:63–104.

Young, C. C. (1951). The Lufeng saurischian fauna. *Palaeontol. Sin., n.s., C* 13:1–96.

Zhang, Y., and Yang, Z. (1994). [*A New Complete Osteology of Prosauropoda in the Lufeng Basin, Yunnan, China:* Jingshanosaurus.] Kunming: Yunnan Publishing House of Science and Technology (in Chinese with English abstract).

PAUL UPCHURCH AND PAUL M. BARRETT

4

The evolution of sauropod feeding mechanisms

Introduction

Sauropods were gigantic, long-necked, herbivorous dinosaurs, which dominated many Jurassic and Cretaceous terrestrial faunas. First appearing in the fossil record during the Early Jurassic, they rapidly achieved a nearly global distribution, with apparent peaks in diversity and abundance in the Kimmeridgian, Hauterivian–Barremian, and Campanian–Maastrichtian (McIntosh 1990; Weishampel 1990; Hunt *et al.* 1994; Barrett 1998).

Early discoveries of poorly preserved sauropod material were interpreted as the remains of marine crocodiles (Owen 1841, 1842). Better preserved specimens (*Cetiosaurus*) from the Middle Jurassic of Oxfordshire, however, allowed Phillips (1871:294) to recognize that these animals were terrestrial dinosaurs which probably ate plants. This discovery was soon overshadowed by spectacular material from the western United States (Cope 1877a,b; Marsh 1877a,b, 1878, 1879), prompting Marsh (1878) to create the new dinosaurian subgroup Sauropoda. The retracted nostrils in the skull of *Diplodocus* (Marsh 1884), and the gigantic size of these animals, led many workers to conclude that sauropods must have been aquatic (Cope 1878; Marsh 1883; Hatcher 1901; Hay 1908). This 'aquatic' hypothesis had an important effect on interpretations of sauropod feeding habits and mechanisms. If such gigantic animals were restricted to food available in or near bodies of water, they probably ate soft aquatic vegetation or invertebrates (Hay 1908; Holland 1924; Haas 1963). This view was seemingly supported by the 'weak and inefficient' masticatory apparatus of most sauropods, especially the 'tooth-comb' of *Diplodocus*, which could have been used for straining items from the water.

Recent studies have provided compelling evidence (from both skeletal structure and trackways) that sauropods were largely terrestrial (Bakker 1971; Coombs 1975; Gillette and Lockley 1989; Dodson 1990). This reinterpretation has led to the suggestion that sauropods were high-browsers that ate vegetation from tree canopies.

Previous studies recognized a twofold division of sauropod feeding mechanisms: (1) nipping/cropping of coarse vegetation by sauropods with heavy spatulate dentitions; (2) 'raking' of softer vegetation by sauropods with more delicate dentitions (Bakker 1971; Norman 1985; Dodson 1990). Whereas these studies provided important insights into sauropod paleoecology, more detailed work has indicated that sauropod feeding was more varied and complex (Barrett and Upchurch 1994, 1995; Calvo 1994a).

Institutional abbreviations in this chapter are: AMNH, American Museum of Natural History, New York; BMNH, The Natural History Museum, London; CMNH, Carnegie Museum of Natural History, Pittsburgh; FMNH, Field Museum, Chicago; CIT, Chengdu Institute of Technology, Chengdu; HMN, Museum für Naturkunde der Humboldt-Universität, Berlin; IVPP, Institute for Vertebrate Paleontology and Paleoanthropology, Academia Sinica, Beijing; LCM, Leicester City Museum and Art Gallery, Leicester; MACN, Museo Argentino de Ciencias Naturales, Buenos Aires; OUMZ, Oxford University Museum of Zoology, Oxford; PMU, Palaeontological Museum, University of Uppsala; USNM, National Museum of Natural History, Washington D.C.; ZDM, Zigong Dinosaur Museum, Zigong.

Materials and methods

The sauropod specimens and sources of data that form the basis for this study are summarized in Tables 4.1 and 4.2. In this analysis, we survey a number of morphological features that reflect aspects of sauropod feeding mechanisms. Tooth shape and position, tooth wear, structure of the jaw joint, orientation of jaw muscles, neck length, and forelimb length are reviewed for each of the major sauropod groups. In addition, data on dental microwear were collected using a scanning electron microscope (SEM). Neck flexibility was assessed on the basis of an examination of intervertebral articulation and the suggestion that shortening of the cervical ribs facilitates greater lateral flexibility. 'Maximum browse heights' are based on the sum of neck and forelimb length; this provides

Table 4.1. *Summary of feeding mechanism characters in sauropod crania*

Taxon	Specimen/source	Dentition	Macrowear	Microwear	Jaw joint	Jaw muscle angle
'Vulcanodontidae'						
Barapasaurus	Jain *et al.* 1975, 1979	Spatulate, denticulate	None	–	?	?
Kotasaurus	Yadagiri 1988	Spatulate, denticulate	None	–	?	?
Euhelopodidae						
Euhelopus	PMU M2983; Mateer and McIntosh 1985	Spatulate	Mesial and distal	–	?	Steep
Mamenchisaurus	Russell and Zheng 1994	Spatulate, denticulate	None	–	Short	Steep
Omeisaurus	IVPP unnumbered, ZDM T5703; He *et al.* 1988	Spatulate, denticulate	Mesial and distal	–	Short?	Steep
Shunosaurus	IVPP 7261–5, ZDM T5401–3; Zhang 1988, Zheng 1991	Spatulate, denticulate	Mesial and distal	–	Short	Steep
'Cetiosauridae'						
Patagosaurus	MACN 124, MACN CH933, MACN CH934; Bonaparte 1986	Spatulate	Mesial and distal	–	?	Steep?
Camarasauridae						
Camarasaurus	CMNH 11338, 21751, AMNH 467; Fiorillo 1991, 1996, Calvo 1994a, Madsen *et al.* 1995	Spatulate	Mesial and distal	Pits (in adults), coarse transverse scratches	Expanded rostrocaudally	Steep
Brachiosauridae						
Brachiosaurus	HMN SII, S66, S116, WJ470; Janensch 1935–36	Cone-chisel/spatulate, some denticles	Apical, subtriangular	–	Short	Steep

Table 4.1 (*cont.*)

Taxon	Specimen/source	Dentition	Macrowear	Microwear	Jaw joint	Jaw muscle angle
Titanosauroidea						
Malawisaurus	Jacobs *et al.* 1993	'Peg'-like	None	–	?	Steep?
Unnamed, Brazil	Kellner 1996	'Peg'-like	Apical, chisel-shaped	-	?	?
Undescribed, Madagascar	FMNH Field No. 93–73/95029, 93-06/95035, 93074	'Peg'-like	Apical, chisel-shaped	Pits, subparallel scratches	?	?
Diplodocidae						
Apatosaurus	CMNH 11162; Berman and McIntosh 1978	'Peg'-like	?	–	?	Shallow
Diplodocus	AMNH 969, CMNH 11161, USNM 2672; McIntosh and Berman 1975, Berman and McIntosh 1978	'Peg'-like	Apical, present on labial surfaces only	Fine subparallel scratches, rostrocaudally oriented	Long	Shallow
Dicraeosauridae						
Dicraeosaurus	HMN dd416, 429; Janensch 1935–36	'Peg'-like	Apical, labial and lingual	–	?	Shallow
Nemegtosauridae						
Nemegtosaurus	Nowinski 1971	'Peg'-like	Apical, labial and lingual	–	Slightly elongate	Less steep

Note: Catalogue numbers are provided for those specimens examined by the authors.

Table 4.2. *Summary of feeding mechanism characters in sauropod postcrania*

Taxon	Specimen/source	Neck/trunk ratio	Cervical ribs	Fl/Hl ratio	Body length	Browse height
'Vulcanodontidae'						
Barapasaurus	Jain *et al.* 1975, 1979	?	?	?	~15 m	?
Kotasaurus	Yadagiri 1988	~1.4	?	0.80	10.5 m	~3.5–4 m
Vulcanodon	Cooper 1984	?	?	0.78	~10 m	?
Euhelopodidae						
Euhelopus	Wiman 1929	2.1	Long	?	?	?
Mamenchisaurus	CIT holotype; Young and Chao 1972	3.5	Long	?	22 m	10–11 m
Omeisaurus	ZDM T5701, T5703–5; He *et al.* 1988	3.8	Long	0.85–0.88	20 m	10–11 m
Shunosaurus	ZDM T5401–2; Zhang 1988	1.35–1.45	Short	0.64	11 m	4.5 m
'Cetiosauridae'						
Cetiosaurus	LCM 468.1968, OUMZ J13605–13690	~1.20	Long?	0.84	~15 m	5 m
Camarasauridae						
Camarasaurus	CMNH 11338, 11939; Osborn and Mook 1921, Gilmore 1925, McIntosh *et al.* 1996	1.25 (juvenile), 1.45 (adult)	Long	0.79 (adult), 0.87 (juvenile)	18 m	~8.5 m
Brachiosauridae						
Brachiosaurus	HMN SII, XV2; Janensch 1950, 1961	2.38	Long?	1.1	25 m	14 m
Titanosauroidea						
Opisthocoelicaudia	Borsuk-Bialynicka 1977	?	?	0.83	?	?

Table 4.2 (*cont.*)

Taxon	Specimen/source	Neck/trunk ratio	Cervical ribs	Fl/Hl ratio	Body length	Browse height
Diplodocidae						
Apatosaurus	CMNH 563, 3018; Gilmore 1936	2.04–2.17	Short	0.70–0.71	23 m	8 m
Barosaurus	Lull 1919, McIntosh 1990	2.5–3?	Short	?	24 m	12 m
Diplodocus	CMNH 84/94, AMNH 223, 655, 5855, USNM 10865; Hatcher 1901, Gilmore 1932	2.13	Short	~0.70	~25 m	9–10 m
Dicraeosauridae						
Amargasaurus	Salgado and Bonaparte 1991	1.36	?	0.70	10 m	4 m
Dicraeosaurus	HMN O, M, m; Janensch 1929, 1936, 1961	1.24	Short	0.64	10 m	4 m

Notes:
Catalogue numbers are provided for those specimens examined by the authors.
Abbreviations: Fl, forelimb length; Hl, hindlimb length.

an estimate for the height of the head if the sauropod stood in a quadrupedal stance with its neck vertical.

The interrelationships of sauropods are based on the cladogram proposed by Upchurch (1995), with the modification that the Titanosauroidea are now considered the sister-group to the Brachiosauridae (Wilson and Sereno 1994; Calvo and Salgado 1995; Salgado, Coria, and Calvo 1997; Upchurch 1998). These relationships, and corresponding classification, are summarized in Figure 4.1.

General comments on sauropod morphology

Reviews of sauropod anatomy can be found in McIntosh (1990), Upchurch (1993, 1994), and Barrett (1998). Some particular features, common to all sauropods, deserve special mention because of their relevance to feeding mechanics.

Sauropod skulls were akinetic. Contacts between the various cranial elements are usually interdigitating sutures or simpler butt/overlap joints held together by ligaments. No transverse or longitudinal hinge joints have been detected. Streptostyly was not possible because the quadrate had fixed contacts with the squamosal, quadratojugal, and pterygoid (Figure 4.2; see also Haas 1963).

In basal Theropoda and Ornithischia, as well as all prosauropods, the rostral end of the skull is narrow and acute in dorsal view. In sauropods, the snout has become wide and rounded in outline. This modification may be related to the cropping of vegetation (see below).

All sauropod skulls, including that of *Shunosaurus*, display some degree of narial retraction (Upchurch 1994, 1995). Most sauropods have laterally facing external nares placed anterodorsal to the antorbital fenestra; in diplodocoids, however, the external nares face upwards and lie anterodorsal to the orbit (Figure 4.2). Bakker (1986) suggested that narial retraction reflects the presence of a fleshy proboscis, as occurs in elephants and tapirs. Such a proboscis could have played a role in food-gathering in sauropods. It seems likely, however, that sauropods lacked the facial musculature that gives rise to the proboscis in certain mammals (Coombs 1975). Alternative explanations for narial retraction include the requirement to elongate the preorbital part of the skull (see below) and/or the need to protect sensitive nasal structures during feeding. At present, however, the significance of narial retraction in sauropods remains obscure.

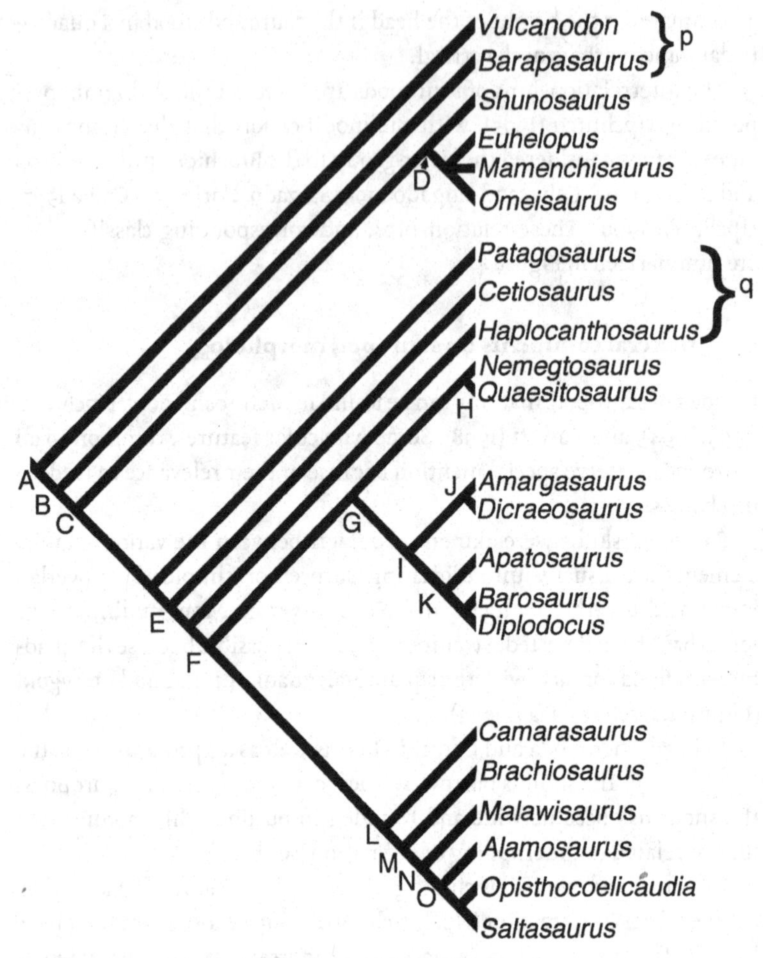

Figure 4.1. Evolutionary relationships and systematic classification of Sauropoda based on Upchurch (1995, 1998) and Salgado *et al.* (1997). Nodes are labelled as follows: A, Sauropoda; B, unnamed node; C, Eusauropoda; D, Euhelopodidae; E, unnamed node; F, Neosauropoda; G, Diplodocoidea; H, Nemegtosauridae; I, unnamed node; J, Dicraeosauridae; K, Diplodocidae; L, Camarasauromorpha; M, Titanosauriformes; N, Titanosauroidea; O, Titanosauridae. Paraphyletic taxa are labelled: p, 'Vulcanodontidae;' q, 'Cetiosauridae.'

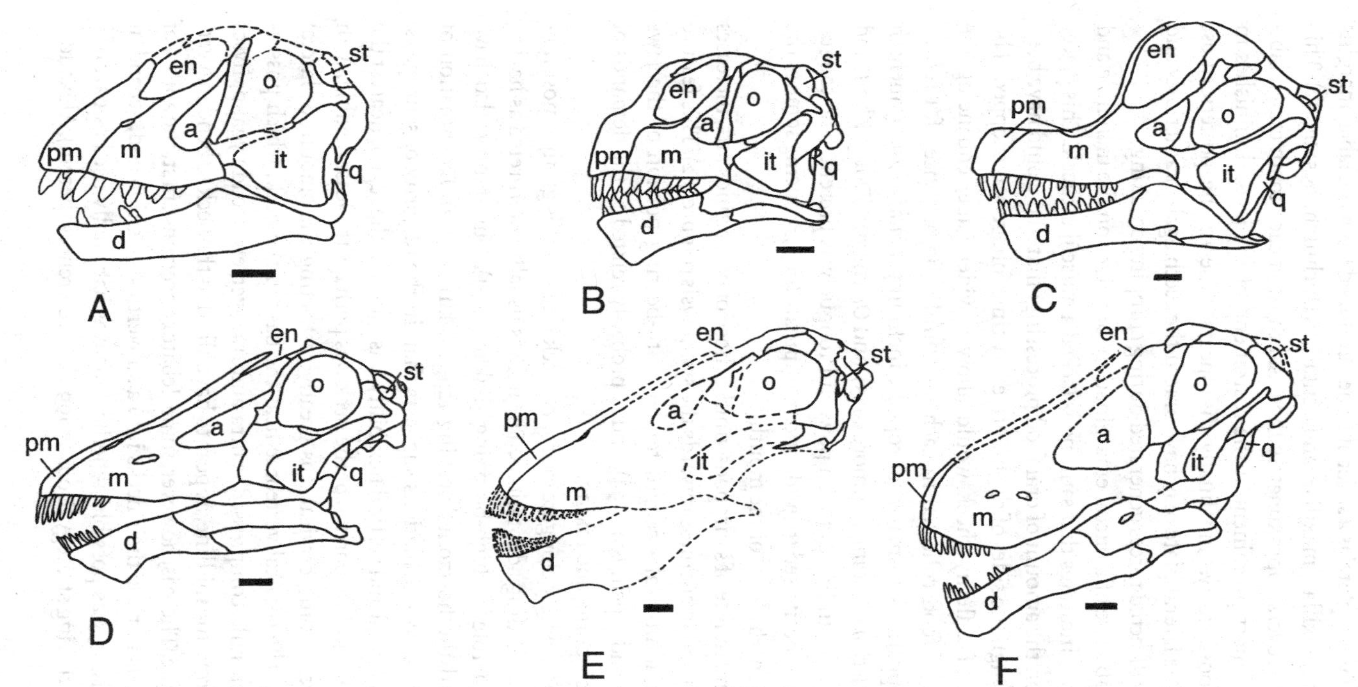

Figure 4.2. Sauropod skulls in lateral view. A, *Euhelopus* (after Mateer and McIntosh 1985); B, *Camarasaurus* (based on CMNH 11338); C, *Brachiosaurus* (after Janensch 1935–36); D, *Diplodocus* (based on CMNH 11161); E, *Dicraeosaurus* (after Janensch 1935–36); F, *Nemegtosaurus* (after Nowinski 1971). Abbreviations: a, antorbital fenestra; d, dentary; en, external naris; it , infratemporal fenestra; m, maxilla; o, orbit; pm, premaxilla; q, quadrate; st, supratemporal fenestra. Scale bars each equal 50 mm.

The 'lateral plate' is a sheet of bone developed from the labial margins of the premaxillae, maxillae, and dentaries (Upchurch 1995: fig. 12). This structure covers approximately the basal third of each tooth crown labially. It is most prominent at the rostral end of the jaws, diminishes in height posteriorly, and finally disappears close to the posteriormost tooth. The lateral plate may have braced the teeth against anteriorly and laterally directed forces generated during cropping or raking.

The absence of a buccal emargination, or ridges on the mandible and upper jaw, indicates that sauropods lacked a muscular cheek. This places a limit on the amount of oral food processing: material would have been lost through the sides of the mouth every time foliage was chopped. The absence of a fleshy cheek would allow a wider gape, enhancing the amount of food collected in each mouthful. The presence of a robust hyoid apparatus in many sauropods, including *Brachiosaurus* (Janensch 1935–36), *Camarasaurus* (Gilmore 1925), and *Omeisaurus* (Dong, Zhou, and Zhang 1983), indicates that a powerful tongue was present. The tongue could have been used to guide and manipulate food, reducing some of the losses from the sides of the mouth.

The mandible of sauropods, unlike those of other dinosaurs, increases in depth and robustness towards the symphysis (Figure 4.2). This feature conveys additional strength to the tooth-bearing portion of the jaws (Barrett and Upchurch 1994), and is probably related to the demands of foliage cropping and raking.

The extremely elongate sauropod neck is one of this group's most fundamental feeding adaptations. The neck is usually interpreted as having played a role in high-browsing (Bakker 1971), but it may also have increased the horizontal feeding range (Martin 1987). Examination of intervertebral articulations suggests that, in the majority of sauropods, greatest lateral and vertical flexibility was found in the region nearest the skull. The middle portion of the neck was perhaps the least flexible, with long overlapping cervical ribs greatly restricting the amount of lateral motion. The cervical vertebrae near the base of the neck, which possess dorsoventrally compressed centra and transversely broad zygapophyses set far from the midline, appear to have allowed the neck to bend in a vertical plane. There is, however, some debate concerning, first, the ability of sauropods to raise their necks into a subvertical position and, second, if neck raising was possible, whether this was habitually used as part of a food-gathering strategy (Dodson 1990 and references therein). Alexander

(1985) discussed the mechanical problems involved in supporting and raising the neck of *Diplodocus*, and concluded that the neck could have been raised if the notch created by neural spine bifurcation was filled by muscle or an elastin ligament, *and* if the area lateral to the neural spines held muscles (as must have been the case in those sauropods that lack neural spine bifurcation). Neck elevation would have been even more likely if the mean density of its tissues had been lowered by air-filled sacs occupying the cervical pleurocoels during life. Hemodynamic considerations have also been used as an argument against sauropod neck raising (Dodson 1990 and references therein). However, mechanisms for coping with such problems, albeit on a smaller scale, are present in extant taxa (Choy and Altman 1992). Therefore, given the nature of the cervical intervertebral joints, and the apparent absence of convincing engineering and physiological counter-arguments, we consider that neck elevation did form part of the food-gathering strategy employed by many sauropods.

The extent to which a 'gastric mill' played a role in the sauropod digestive system is controversial. Calvo (1994b) claimed that the association between gastroliths and sauropod skeletal material was only reliable in the case of two specimens from Argentina (aff. *Rebbachisaurus*). Christiansen (1996), however, cites the presence of gastroliths in close association with *Apatosaurus, Barosaurus, Camarasaurus alenquerensis, Dicraeosaurus, Diplodocus* and *Seismosaurus*. A gastric mill is also present in the abdominal region of an undescribed brachiosaurid from the Isle of Wight (S. Hutt, pers. comm.). Calvo (1994b) predicted that sauropods that utilized little or no oral processing (such as brachiosaurids and diplodocoids) would be more likely to need a gastric mill than those with substantial oral processing (such as camarasaurids). The 'known' distribution of gastroliths is consistent with Calvo's prediction, since all of the sauropods listed above are either brachiosaurids or diplodocoids, with the possible exception of *Camarasaurus alenquerensis*. The taxonomic status of the latter, from the Upper Jurassic of Portugal, is problematic, and this taxon possesses at least one diplodocoid synapomorphy (Upchurch 1995). Therefore, whereas microbial fermentation (facilitated by a long digestive tract, slow passage rates, and presumably low mass-specific basal metabolic rates) may have satisfied a sauropod's energy requirements (Farlow 1987), gastric mills were probably used to enable mechanical breakdown of food in at least some of the forms that were not capable of oral processing.

Sauropod feeding-related characters

'Vulcanodontidae'

The 'Vulcanodontidae' are a paraphyletic assemblage of basal sauropods from the Early Jurassic (McIntosh 1990; Upchurch 1995). *Vulcanodon*, *Barapasaurus*, and especially *Kotasaurus* are known from reasonably complete postcrania, but cranial material is extremely rare.

Skull A pair of partial lower jaws from the Lower Lufeng Formation (Hettangian–Pliensbachian: Luo and Wu 1995) of the Wuding Basin, Yunnan, have been referred to *Kunmingosaurus* (Dong 1992). These specimens possess a lateral plate and a ventrally expanded symphysis. Even if the assignment of these mandibular remains to *Kunmingosaurus* is incorrect, they mark the occurrence of some 'typical' sauropod feeding adaptations early in the Jurassic.

Dentition A tooth crown (IVPP, unnumbered) from the Lower Lufeng Formation has been referred to *Kunmingosaurus* (Z. Dong, pers. comm.). It is not clear whether this tooth was derived from the upper or lower jaw, and it is thus impossible to distinguish the mesial and distal margins of the crown. This specimen retains several plesiomorphic character-states, such as a labiolingually narrow base and only moderate mesiodistal expansion of the crown. It also possesses derived states present in other sauropods, including a shallowly concave lingual surface, a median ridge within the lingual concavity, and a groove on the labial surface which runs parallel to either the mesial or distal margin. The apex of the tooth lacks any distal curvature, although it is lingually inclined as in most sauropods. Both margins of the crown are coarsely serrated, with up to nine denticles on one, and at least three on the other (damaged) margin.

Some finely serrated recurved teeth were found in association with the skeleton of *Vulcanodon*, but these are likely to have been shed by scavengers (Cooper 1984). The tooth crowns of *Barapasaurus* and *Kotasaurus* more closely resemble the typical spatulate type found in euhelopodids, cetiosaurids and camarasaurids: the crowns have become more robust, with labiolingually thickened bases; the lingual surface is strongly concave; and, on the labial face, grooves extend parallel to the mesial and distal margins (Jain *et al.* 1979; Yadagiri 1988). As in *Kunmingosaurus*, the tooth crowns of *Barapasaurus* and *Kotasaurus* retain denticles.

Tooth wear No wear has been observed on any vulcanodontid tooth.

Jaw joint and musculature No information available at present.

Postcrania The neck of *Kotasaurus* consists of 12 or 13 cervical vertebrae and was approximately 2.3 m long (Yadagiri 1988). The length of the cervical ribs is not known in this form.

The forelimb/hindlimb ratios in *Vulcanodon* and *Kotasaurus* (Table 4.2) should be treated with caution: several limb-bones of *Vulcanodon* are only partially preserved, and Yadagiri (1988) used *Vulcanodon* as a model to estimate limb proportions in *Kotasaurus*. Nevertheless, current data indicate that vulcanodontids possessed long forelimbs compared with non-sauropods (where the forelimb/hindlimb ratio is usually 0.6 or less). Quadrupedality may be partially linked to this increase in forelimb length, but other dinosaurian quadrupeds (stegosaurs, ankylosaurs, and ceratopians) have forelimb/hindlimb ratios lower than those of sauropods (Barrett 1998). Therefore, we interpret forelimb elongation, and the concomitant elevation of the anterior part of the thorax, as a reflection of the adaptation towards higher browsing (see below).

Feeding mechanism Compared with prosauropods and other 'basal' dinosaurs, vulcanodontids have increased the lengths of the neck and forelimbs, suggesting an early adaptation to browsing at a greater height above ground. Relative to many more derived sauropods, however, vulcanodontids were only small or medium-sized and possessed short necks; most early sauropods probably browsed no higher than about 3–4 m (Table 4.2). *Barapasaurus*, with its greater body size (with a length of about 15 m), may have been able to browse at higher levels. The little that is known about vulcanodontid skulls suggests that they fed by cropping vegetation. The absence of tooth wear implies that mechanical breakdown of food did not occur in the mouth, although more complete material may contradict this hypothesis.

Euhelopodidae

The Euhelopodidae represent a monophyletic group of sauropods endemic to the Jurassic and Early Cretaceous of China (Figure 4.1)..

Skull Well-preserved skull material is known for *Euhelopus*, *Omeisaurus* and *Shunosaurus*, and information on the palate and mandible of *Mamenchisaurus* is now available (Russell and Zheng 1994). These skulls possess the typical sauropod features described above, and, in many

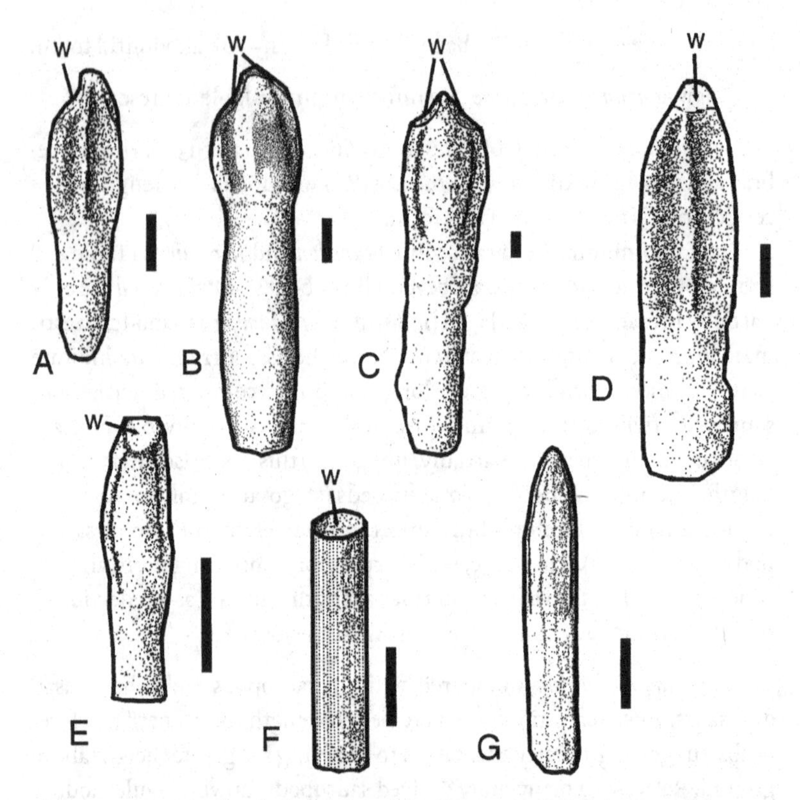

Figure 4.3. Sauropod teeth. A, *Euhelopus* (after Wiman 1929); B, *Patagosaurus* (after Bonaparte 1986); C, *Camarasaurus* (after Osborn and Mook 1921); D, *Brachiosaurus* (after Janensch 1935–36); E, undescribed titanosaurid; F, *Diplodocus* (based on CMNH 11161); G, *Dicraeosaurus* (after Janensch 1935–36). All teeth are in labial view except A and D, which are in lingual view. 'W' marks the position of wear facets. Scale bars each equal 10 mm.

respects, are intermediate in shape between the short, high skulls of camarasaurids and brachiosaurids and the long, low skulls of diplodocoids (Figure 4.2).

Dentition Most euhelopodids have spatulate teeth (Figure 4.3). In *Shunosaurus*, *Euhelopus*, and *Mamenchisaurus*, the tooth crowns are slender, with moderate mesiodistal expansion. In contrast, the teeth referred to *Omeisaurus* are much more robust, with shorter crowns and greater mesiodistal expansion. The denticles in *Shunosaurus* are small and confined to the area immediately adjacent to the apex of the tooth crown, whereas they are large and continue to a point about one-third of the way down the crown in *Omeisaurus*. Denticles are absent in *Euhelopus*.

Tooth wear The teeth of *Euhelopus* possess prominent, high-angle wear facets on the mesial and distal margins, extending from the apex to the point where the crown attains its greatest mesiodistal width (e.g., PMU M2983; Figure 4.3). These facets have a characteristic 'step-flush' appearance: a step is formed between the enamel and dentine at the base of the wear facet, but this step decreases in height and finally disappears completely close to the apex of the tooth crown. This type of wear is produced by tooth–tooth contact. As the opposing tooth rows interlock during jaw closure, the margins of the opposing crowns shear past each other. The areas immediately adjacent to the apices suffer the highest amount of wear, as these are the first to be worn away following eruption and continue to be worn throughout the functional life of the tooth. As a result, the enamel–dentine interface is worn down to a flush contact close to the apex. The step pattern, caused by differential wear of enamel and dentine, persists on the 'lower' part of the wear facet because this area is worn later in the functional life of the tooth. Similar wear is also present on several teeth of *Shunosaurus* and *Omeisaurus*. Some teeth of *Euhelopus* and *Omeisaurus* are very heavily worn, resulting in 'shoulders' on the crown margins (Figure 4.3). Small terminal wear facets may be present on the teeth of *Shunosaurus* (Zhang 1988).

It is difficult to assess the distribution of wear facets along the tooth rows of euhelopodids. In the most complete skull of *Shunosaurus* (ZDM T5403), the lingual surface of the upper tooth row is obscured by the mandible, and the entire lower tooth row is hidden by the upper teeth. The most complete specimens of *Euhelopus* and *Omeisaurus* do not preserve entire tooth rows. Wear is absent on all of the teeth of *Mamenchisaurus sinocanadorum* (Russell and Zheng 1994).

No information on the microwear of euhelopodid teeth is available at present.

Jaw joint and musculature The jaw joint of euhelopodids is short and limited anteriorly and posteriorly by low transverse ridges on the articular. Some caution is necessary in the interpretation of the glenoid fossa in *Omeisaurus* (ZDM T5703) because its shape may have been affected by preparation. The jaw joint of *Shunosaurus* is almost level with the tooth row, whereas those of *Omeisaurus*, *Euhelopus* and *Mamenchisaurus* are offset ventrally.

The structure of the palate and temporal region suggests that the adductor jaw musculature of euhelopodids was steeply inclined relative to the long axis of the skull. This, combined with the structure of the jaw joint and tooth wear, implies that jaw motion was strictly orthal.

Postcrania Euhelopodids can be divided into short-necked (*Shunosaurus* and *Datousaurus*) and long-necked forms (*Euhelopus*, *Mamenchisaurus*, and *Omeisaurus*). The former condition, in which there are 12 or 13 short cervicals, is the plesiomorphic character-state (Upchurch 1995). In Shunosaurus, the cervical ribs are short, only extending posteriorly for the distance of a single cervical vertebra, and do not come into close contact with each other (PU, pers. obs.). This, combined with the shape and horizontal orientation of the cervical zygapophyseal articular facets, suggests that the neck was capable of more lateral flexibility than in most other sauropods. In long-necked euhelopodids, neck elongation has been achieved by lengthening individual cervicals, and by increasing the number of cervical vertebrae from 13 to 17 (Upchurch 1994). The cervical ribs are greatly elongated (often over 3 m long in *Mamenchisaurus*), with the rib shafts forming overlapping 'bundles' (Young and Chao 1972; He, Li, and Cai 1988; Russell and Zheng 1994).

Incomplete preservation of the known specimens means that forelimb/hindlimb ratios are not available for *Euhelopus* and *Mamenchisaurus*. *Shunosaurus* has relatively short forelimbs, whereas those of *Omeisaurus* are approximately 85% of hindlimb length (Table 4.2).

Feeding mechanism Euhelopodids utilized a simple orthal jaw action. The presence of well-developed mesial and distal wear facets indicates a powerful shearing action, with more oral processing than in vulcanodontids and prosauropods. Small body size, a short neck, abbreviated cervical ribs, the nature of the cervical zygapophyseal articulations, and the shortened forelimbs, all imply that *Shunosaurus* was a low-level browser, obtaining food over a wide area by moving the neck laterally as well as vertically. The long-necked euhelopodids were probably specialized for high-browsing, obtaining foliage at heights perhaps as great as 10 m to 11 m.

'Cetiosauridae'

The taxa included within this family vary according to different authors (compare McIntosh 1990 and Upchurch 1995). Here, the 'Cetiosauridae' form an assemblage of Middle and possibly Late Jurassic sauropods, which is paraphyletic with respect to the Neosauropoda (Figure 4.1).

Skull No complete cetiosaurid skull has been described, although well preserved cranial material is available for '*Cetiosaurus mogrebiensis*' (McIntosh 1990). A few isolated teeth from the Middle Jurassic of England have been referred to *Cetiosaurus* (e.g., BMNH R3377), but this referral is dubious. Therefore, this study is based on the partial snout, dentary, and isolated teeth of *Patagosaurus* (Bonaparte 1986). These elements suggest that the cetiosaurid skull resembled those of euhelopodids and *Camarasaurus*.

Dentition The teeth of *Patagosaurus* have mesiodistally expanded spatulate crowns. Denticles are absent.

Tooth wear In *Patagosaurus*, there are well-developed mesial and distal wear facets (MACN 124, MACN CH933, MACN CH934; Bonaparte 1986), resembling those found in euhelopodids (Figure 4.3). Heavily worn teeth develop 'shoulders.'

No work on the microwear of cetiosaurid teeth has been reported.

Jaw joint and musculature No information on the structure of the jaw joint is currently available. Our limited knowledge of the cetiosaurid skull suggests that jaw adductor muscles were steeply inclined, but this requires confirmation by better preserved specimens.

Postcrania *Cetiosaurus* (LCM 468.1968) has 12 or 13 short cervicals, resulting in a relatively short neck (Table 4.2). The cervical ribs of this specimen are fragmentary, but probably had long overlapping shafts. Material not incorporated into the mounted skeleton of LCM 468.1968 includes many long, slender portions of distal cervical rib shafts. Furthermore, the preserved proximal ends of the ribs possess a long anterior process, which is typically associated with a long distal shaft in sauropods. The structure of the neck in *Patagosaurus* is less well known, but probably closely resembled that in *Cetiosaurus*.

The forelimb/hindlimb ratio in *Cetiosaurus* is approximately 0.84.

Feeding mechanism Our limited knowledge of cetiosaurid skulls indicates a jaw action similar to that in euhelopodids. The moderate body size and neck length, restricted lateral movement of the neck, and relatively long forelimbs, suggest that cetiosaurids were high-browsers, perhaps obtaining food 5–6 m above ground level.

Camarasauridae

Only *Camarasaurus* is known in sufficient detail to provide information on the feeding adaptations of this group.

Skull The cranial structure of *Camarasaurus* is extremely well known (Madsen, McIntosh, and Berman 1995). The skull is robust, somewhat taller (relative to its anteroposterior length) than those of euhelopodids (Figure 4.2).

Dentition The teeth of *Camarasaurus* possess mesiodistally expanded spatulate crowns. Denticles are normally absent, even in unworn or unerupted teeth. The apices of the crowns are slightly curved distally and lingually.

Tooth wear Wear facets are present, but their shape and size depend in part on the position of the tooth along the tooth row (Carey and Madsen 1972). For example, wear is heaviest on more anteriorly placed teeth, especially those in the upper jaw. *Camarasaurus* teeth frequently display mesial and distal wear facets similar to, but usually larger than, those found in euhelopodids and cetiosaurids (Figure 4.3). The formation of 'shoulders' is also very common. Small terminal wear facets occur occasionally.

Three studies have examined dental microwear in *Camarasaurus* (Fiorillo 1991, 1996; Calvo 1994a). Fiorillo (1991) described coarse scratches and pitting on the wear facets of teeth in adult *Camarasaurus*, which he ascribed either to low-level browsing (more grit is present in the lower levels of tree canopies, and the amount of grit in the diet has a strong influence on tooth microwear) or feeding on coarse vegetation. Fiorillo (1996) examined the microwear on teeth belonging to a juvenile specimen of *Camarasaurus*, and found fine scratches and no pits. This result indicates that the size of microwear scratches and pitting is not height dependent (at least in sauropods), since the juvenile *Camarasaurus* would have been browsing at heights lower than those exploited by adults. These differences in microwear might indicate ontogenetic variation in diet. Calvo (1994a) observed pits and scratches, and noted that the majority of the latter were oriented parallel to the labiolingual axis of the tooth.

Jaw joint and musculature The jaw articulation of *Camarasaurus* is offset ventrally. The shallow concave articular surface is anteroposteriorly expanded and slopes downwards anteromedially. This articular surface is significantly longer anteroposteriorly than the distal condyle of the quadrate, indicating that propalinal motions were possible (Calvo 1994a; Barrett and Upchurch 1995). White (1958) and Calvo (1994a) also noted the lateromedial expansion of the articular surface, and argued that *Camara-*

saurus would have been capable of a transverse power stroke. Lateral movements of the mandibular rami, however, require the presence of a complex jaw musculature such as the masseter 'sling' of mammals. A system equivalent to a masseter sling has never been found in any sauropsid, and the cranial structure of *Camarasaurus* does not support the presence of such a structure. Furthermore, the size and position of the teeth in *Camarasaurus* would make transverse motions of the lower jaw impossible when the jaws were close to occlusion. Therefore, the production of a transverse power stroke in *Camarasaurus* is considered highly unlikely. The orientation of the adductor jaw musculature suggests that jaw motion was essentially orthal, with some propaliny possible if the internal and external adductor muscles were contracted 'out of phase.' It is not clear yet whether the longitudinal component in jaw motion was proal, palinal, or some combination of the two.

Postcrania The neck of *Camarasaurus* is relatively short compared with overall body size (Table 4.2). The ribs of the middle cervicals possess long, overlapping distal shafts (Gilmore 1925).

Camarasaurus possessed relatively long forelimbs (Table 4.2). Some of the observed range in forelimb/hindlimb ratio (i.e., 0.79 in adults to 0.87 in juveniles) may reflect ontogenetic variation, but it is also possible that the adult values are inaccurate because of the incorrect association of disarticulated material (Osborn and Mook 1921).

Feeding mechanism The jaw action of *Camarasaurus* was apparently more complex than those employed by euhelopodids and cetiosaurids. The structure of the jaw joint, orientation of the adductor jaw muscles, and tooth wear are consistent with an orthal shearing action and a moderate amount of propaliny. The presence of labiolingually directed microwear scratches is unexpected, but does not necessarily indicate a transverse motion of the jaws. A food-gathering technique that involved tugging/raking of foliage from branches might cause material to be pulled transversely across the wear facets. Alternatively, some food might have been dragged transversely across the wear facets as the upper and lower teeth sheared past each other. Perhaps raking was employed during the initial phase of each food-gathering episode, with orthal shearing and some propalinal motion being used to finally sever material from the rest of the branch. *Camarasaurus* may have fed on coarser vegetation than other sauropods, a view that is consistent with the greater amount of oral processing implied by dental macrowear. The neck of *Camarasaurus* is not particularly elongate, but the large body size of this animal allowed browsing at heights of 8–9 m above ground level.

Brachiosauridae

Although several genera have been referred to the Brachiosauridae (McIntosh 1990; Upchurch 1994, 1995), only *Brachiosaurus* itself is known adequately from both skull and postcranial material.

Skull The skull of *Brachiosaurus* is relatively tall, with a steeply arching internarial bar. The region rostral to the external nares forms a long, flattened 'muzzle,' which is much more prominent than in any other sauropod (Figure 4.2).

Dentition Brachiosaurid teeth are spatulate, but the crowns are only slightly wider mesiodistally than the roots. This more slender form of spatulate tooth has been termed a 'cone-chisel' tooth by Calvo (1994a). A few denticles are sometimes present near the apex of unworn teeth (HMN WJ470; Janensch 1935–36).

Tooth wear In well-preserved jaws of *Brachiosaurus*, a few of the teeth possess typical mesial and distal wear facets or even 'shoulders.' The majority of worn teeth, however, bears a large subtriangular facet situated at the crown apex (Figure 4.3). These facets face ventrolingually and dorsolabially on the upper and lower teeth, respectively. The prevalence of these apical facets, and the reduced mesial and distal wear, is the result of a modification in the relative positions of the tooth rows: instead of 'interdigitating,' the upper and lower teeth are aligned, allowing a more precise shearing bite (Calvo 1994a).

No microwear studies of teeth of *Brachiosaurus* have been reported.

Jaw joint and musculature A low ridge extends transversely across the posterior part of the articular glenoid, greatly restricting fore-and-aft motion of the lower jaw. The adductor jaw musculature, especially in the temporal region, would have been steeply inclined with respect to the long axis of the lower jaw, implying a simple orthal jaw action.

Postcrania Although *Brachiosaurus* possesses only the plesiomorphic number of cervical vertebrae (12 or 13), elongation of individual cervicals produces a long neck (Table 4.2). Janensch (1950) and Paul (1988) reconstructed the cervical ribs of *Brachiosaurus* as relatively short: each rib extending backwards to overlap the anterior portion of the succeeding vertebra. The cervical ribs of the best preserved skeletons of *Brachiosaurus*, however, are badly broken. Given that each rib possesses a long anterior projection, it is likely that the distal shafts were longer than existing reconstructions would imply (see 'Cetiosauridae').

The forelimbs of *Brachiosaurus* are extremely long (Table 4.2), resulting in a substantial elevation of the cranial portion of the thorax.

Feeding mechanism *Brachiosaurus* seems to have developed a more precise shearing bite than other 'broad-toothed' sauropods. Therefore, food-gathering would have involved more 'cutting' through foliage than tearing and ripping. Reduction in the prominence of mesial and distal wear facets implies less oral processing. The elongation of the neck and forelimbs, and the extremely large body size, indicate that *Brachiosaurus* was specialized for browsing at heights somewhat greater than those achieved by other sauropods.

Titanosauroidea

The Titanosauroidea are represented by a relatively large number of valid genera. However, paradoxically, fragmentary preservation of most known specimens means that we know less about the anatomy and relationships of this group than any other sauropod clade (McIntosh 1990).

Skull No complete skull is available for any titanosauroid, though a few cranial fragments (Table 4.1) offer tantalizing clues to its form and function. The best preserved tooth-bearing elements are the premaxilla and dentary of *Malawisaurus* (Jacobs et al. 1993) and possibly a maxilla from the Upper Cretaceous of India (Huene and Matley 1933:23–24). Until recently, reconstructions of the titanosauroid skull resembled the long, low, crania of diplodocoids (Huene 1929). This view was based on material assigned to *Antarctosaurus wichmannianus*, which is more likely to have belonged to a diplodocoid (Jacobs et al. 1993; Upchurch 1999). The premaxilla of *Malawisaurus* indicates that the titanosauroid skull was actually short and high, as in *Camarasaurus* and *Brachiosaurus*. Such an interpretation is also consistent with the revised phylogenetic position of this clade (Figure 4.1).

Dentition Titanosauroid teeth resemble those of *Brachiosaurus* in being relatively slender and possessing nearly parallel mesial and distal margins, but they have developed additional apomorphies such as the absence of the lingual concavity and a decrease in relative size (Figure 4.3). The lingual surface of the crown may be mildly or strongly convex, producing an elliptical or cylindrical horizontal cross-section, respectively.

The crown terminates in a bluntly rounded subtriangular apex, which is often curved lingually.

Tooth wear Wear facets are present on the teeth of *Titanosaurus* (Huene 1929) and undescribed titanosaurs from Madagascar (C. A. Forster, pers. comm.), Brazil (Kellner 1996) and Morocco (Sereno et al. 1996). These teeth, although found in association with titanosauroid material, were not recovered in contact with jaw elements. The mesial and distal high-angle wear facets present on spatulate teeth have not been observed on any titanosauroid tooth. Instead, there is a high-angle wear facet at the apex of each crown, resulting in a 'chisel'-shaped tip (Figure 4.3). Assuming these isolated teeth curve lingually (as in most sauropods, but see 'Nemegtosauridae'), it appears that wear facets are found on the labial surface of some teeth and the lingual surface of others. This could be interpreted as resulting from a precise occlusion between upper and lower tooth crowns, producing wear on the labial surfaces of lower teeth and the lingual surfaces of upper teeth.

The microwear on the teeth of the undescribed Madagascan titanosaurid includes coarse scratches (30–40 μm wide) extending parallel to the long axis of the crown (Figure 4.4). These scratches often extend along the entire length of the wear facet. There are also randomly distributed pits. This wear pattern could have been caused by durable food particles, or grit, crushed between the teeth during occlusion. If the lingual and labial wear facets occurred on the upper and lower teeth, respectively, the scratches could have been caused by tooth–tooth contact.

Jaw joint and musculature No information on the titanosauroid jaw joint is available at present. If the skull was short and tall, the adductor jaw musculature would have been steeply inclined.

Postcrania Information on the neck structure of titanosauroids is mainly derived from disarticulated materials, although nearly complete cervical series are known from Brazil (Powell 1987). *Saltasaurus* has relatively short cervicals in comparison with those of the unnamed Brazilian form (Powell 1992), suggesting that relative neck length varied between different titanosauroid groups. The cervical ribs are partially preserved in the Brazilian titanosaur; the anterior processes are very prominent and at least some of the distal shafts were elongate (Powell 1987: pl. 1).

The best preserved articulated postcranial skeleton of a titanosauroid belongs to *Opisthocoelicaudia* (Borsuk-Bialynicka 1977; Upchurch 1995). This specimen, which possesses virtually complete articulated limbs,

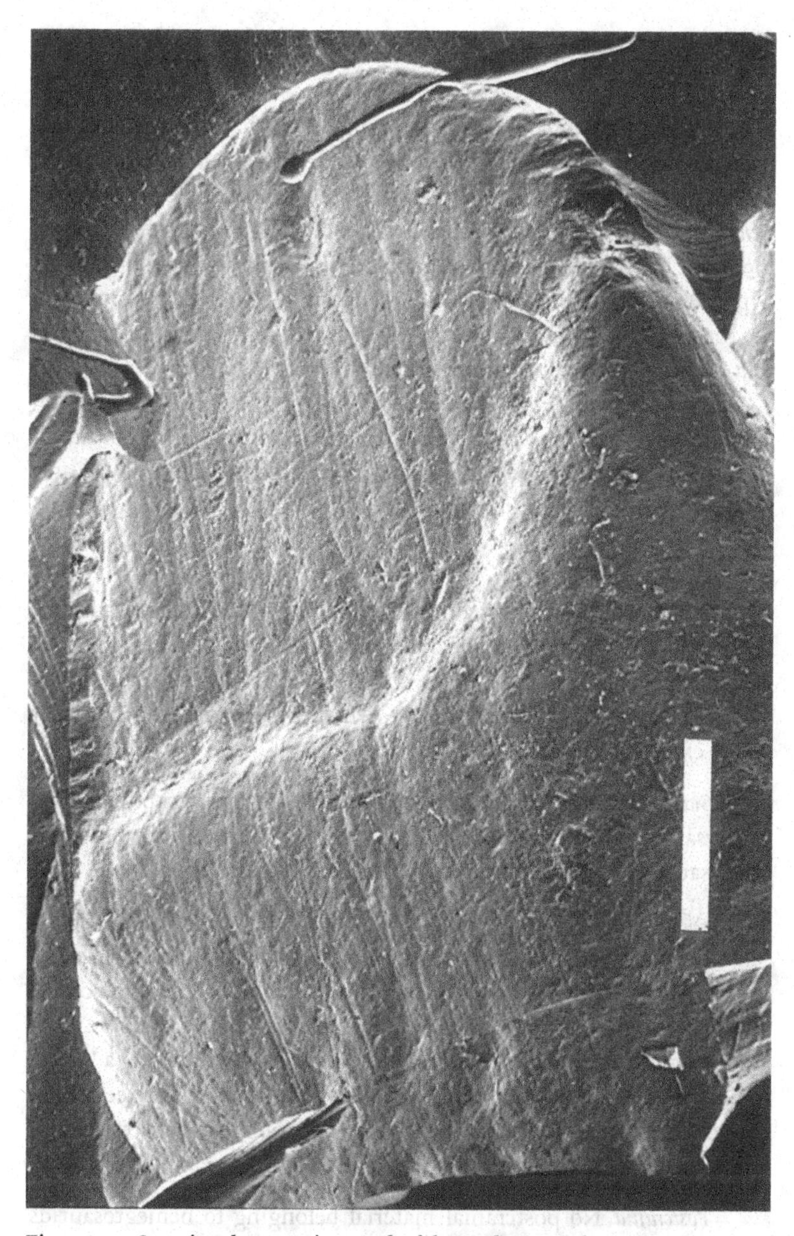

Figure 4.4. Scanning electron micrograph of the tooth crown of an unnamed titanosaurid from Madagascar (FMNH field no. 93–73/95029). This picture was obtained from a high-resolution peel of the tooth. The micrograph is oriented with the apex of the crown towards the top of the page. Although most of the microwear is located on the surface of the wear facet (rhomboidal area in the top half of the picture), some wear striae are also present on the 'unworn' crown surface. Scale bar equals 1 mm.

demonstrates that at least some titanosauroids had relatively long forelimbs (Table 4.2).

Feeding mechanism Calvo (1994a) suggested that titanosaurids utilized orthal jaw motions with precise tooth–tooth contact in order to produce a shearing bite. Although this view was based on a study of *Nemegtosaurus* (which is now considered a diplodocoid; Upchurch 1995, 1999), the data obtained from undoubted titanosauroids support Calvo's hypothesis. Maximum browse heights cannot be estimated for most titanosauroids, although the structure of the neck and forelimbs suggests that some forms fed at 'intermediate' heights, approximately 6–8 m above ground level. Given the observed variation in cervical elongation and body size, and the absence of information concerning cranial structure, it is likely that future discoveries will reveal considerable diversity in titanosauroid feeding mechanics.

Diplodocoidea

The Diplodocoidea comprise three families, the Nemegtosauridae, Dicraeosauridae, and Diplodocidae (Upchurch 1995). The skulls of diplodocoids share a number of characteristics relevant to inferring feeding (Barrett and Upchurch 1994), which will be discussed together before examining each family individually.

Skull The skull is long and low, with an elongate preorbital region. In dorsal view, the mandible is subrectangular in outline, whereas in other sauropods the dentaries curve gradually round towards the symphysis (Upchurch 1998: fig. 5).

Dentition The tooth crowns are 'peg'-like in all diplodocoids. The teeth of nemegtosaurids and dicraeosaurids resemble those of titanosauroids, whereas diplodocids possess a unique slender tooth form. The teeth are restricted to the extreme rostral ends of the jaws, producing a 'toothcomb.' As in Titanosauriformes, the relative positions of the teeth have been modified in diplodocoids so that one tooth on the upper jaw contacts one or two teeth on the lower jaw (Salgado and Calvo 1997; Upchurch 1999).

Postcrania No postcranial material belonging to nemegtosaurids has been recovered to date. The cervical ribs of dicraeosaurids and diplodocids are short and do not form overlapping 'bundles.' The forelimbs in both taxa are shorter than those of other sauropods (Table 4.2).

Nemegtosauridae

Nemegtosaurids are a poorly known group represented by two isolated skulls from the Upper Cretaceous of Mongolia.

Skull The nemegtosaurid skull shares several derived character-states with dicraeosaurids and diplodocids (see above), but also retains some plesiomorphic features. For example, the snout is elongated antero-posteriorly and is more rectangular in dorsal view, but the posterior part of the skull has undergone less modification than in other diplodocoids (Upchurch 1999).

Dentition In *Nemegtosaurus* (Nowinski 1971), and probably *Quaesitosaurus* (Kurzanov and Bannikov 1983), the teeth are not strongly procumbent. The upper teeth curve lingually, but the lower ones are highly unusual because they curve slightly labially.

Tooth wear No tooth wear was reported in the description of *Quaesitosaurus* (Kurzanov and Bannikov 1983), but extensive wear is present in *Nemegtosaurus* (Nowinski 1971). Apical wear facets occur on the labial and lingual surfaces of lower and upper teeth, respectively. There is also a small amount of wear on the mesial and distal margins of some crowns.

No studies of the microwear on nemegtosaurid teeth have been reported.

Jaw joint and musculature The glenoid area, on the dorsal surface of the articular, is slightly elongated anteroposteriorly (Nowinski 1971). Elongation of the snout and the possibility that the quadrate sloped a little anteroventrally, suggest that the adductor jaw muscles were oriented at a shallower angle to the long axis of the skull. These features indicate that at least a moderate amount of propalinal movement could have been achieved.

Feeding mechanism The tooth wear observed in nemegtosaurids suggests that a shearing bite was employed. The presence of mesial and distal wear could reflect some oral processing, or a less precise tooth-to-tooth contact than that found in titanosauroids and dicraeosaurids. The role played by propalinal movements is not entirely clear, although it may have widened the gape and facilitated correct relative positioning of the upper and lower tooth rows during the bite phase.

Dicraeosauridae

The Dicraeosauridae are represented by two genera, *Dicraeosaurus* and *Amargasaurus*, both of which are known from partial skull and postcranial material.

Skull The skull material assigned to *Dicraeosaurus* (Janensch 1935–36) and *Amargasaurus* (Salgado and Bonaparte 1991; Salgado and Calvo 1992) is fragmentary. Nevertheless, the available specimens indicate that the dicraeosaurid skull was similar in proportions and structure to those of diplodocids. The snout region is elongate, and the quadrates probably sloped anteroventrally (see below).

Dentition The slender 'peg'-like teeth of *Dicraeosaurus* (Figure 4.3) are slightly compressed labiolingually, with a gently convex lingual surface. Both labial and lingual surfaces bear grooves paralleling each of the mesial and distal margins. The crown tapers towards the apex and curves lingually. The mesial and distal margins of the crown form acute edges.

Tooth wear None of the worn teeth assigned to *Dicraeosaurus* has been found *in situ*, making it impossible to determine the true orientation of wear facets. Assuming that all of the isolated teeth curve lingually (but see Nemegtosauridae), it appears that small subtriangular wear facets are present on the labial surfaces of some crown apices and the lingual surfaces of others. This wear pattern is very similar to that observed in titanosauroids and was probably formed in the same way.

No microwear studies of dicraeosaurid teeth have been reported.

Jaw joint and musculature The jaw joint is not preserved in either *Dicraeosaurus* or *Amargasaurus*. The phylogenetic relationships of dicraeosaurids, however, suggest that the glenoid was probably at least somewhat anteroposteriorly elongated. It is likely that the adductor jaw musculature was oriented at a low angle to the long axis of the skull: the preorbital region is elongated and the anteroventral slope of the quadrate is implied by the very long, anteriorly directed basipterygoid processes (Salgado and Calvo 1997; Upchurch 1999). Propaliny may, therefore, have played an important role in the jaw action.

Postcrania Dicraeosaurids possess 11 or 12 short cervical vertebrae, producing a rather short neck (Table 4.2). The cervical vertebrae bear extremely tall neural spines (Janensch 1936; Salgado and Bonaparte 1991), which would have impeded raising the head far above the level of the shoulders.

Feeding mechanism The tooth wear indicates that dicraeosaurids probably used a precision shear bite. The role played by propalinal movements is not fully understood (see Nemegtosauridae above). The short neck and forelimbs restricted vertical and increased lateral motion of the neck, and the comparatively small body size (Table 4.2) suggest that dicraeosaurids were specialized low-level browsers. Rearing up into a bipedal stance could have increased browse height, but the absence of pleurocoels in the cervical and dorsal centra seems to indicate that strength of the vertebral column was more important than lightness.

Diplodocidae

The Diplodocidae contain a number of well-known sauropods, most of which come from the Upper Jurassic of North America.

Skull The skull of *Diplodocus longus* (CMNH 11161) is particularly well preserved and forms the basis for much of the following discussion. Although fragmentary, the cranial material assigned to *Apatosaurus* closely resembles that of *Diplodocus* (Berman and McIntosh 1978).

In *Diplodocus,* the ventral margin of the maxilla and dorsal margin of the dentary are formed into thin sheets of bone that overlap each other when the jaws are fully or nearly closed. This combination of structures, which has not been observed in other sauropods, could be regarded as a bony equivalent of a fleshy cheek: it would have prevented loss of food from the lateral margins of the mouth (see 'Feeding mechanism' below). The quadrate slopes anteroventrally, so that the temporal region is situated above and behind the jaw joint (Figure 4.2). As in dicraeosaurids, the occipital condyle is directed ventrally rather than posteroventrally relative to the long axis of the skull (Berman and McIntosh 1978). This orientation could be regarded as an inevitable by-product of other modifications to the posterior part of the skull (Salgado and Calvo 1997; Upchurch 1999), but might also be related to an alteration in the habitual posture of the head (relative to the neck) during feeding. There is a long retroarticular process in *Diplodocus.*

Dentition The teeth of *Diplodocus* are highly unusual compared with those of other sauropods. The long slender crowns are elliptical in cross-section and lack any trace of labial and lingual grooves. The mesial and distal margins are parallel to each other for virtually the entire length of the tooth, rapidly converging at the apex to form a blunt triangular point. There is little lingual curvature of the crowns. The teeth are distinctly 'procumbent' (i.e., their long axes are directed rostroventrally and rostrodorsally in upper and lower teeth respectively). The upper teeth are a little longer, and somewhat more robust, than the lower teeth.

Tooth wear The teeth of *Diplodocus* frequently display heavy wear (contra Haas 1963). No wear has been observed on the mesial and distal margins, but a prominent facet is often present at the apex (Figure 4.3). Wear is somewhat heavier on the upper teeth than on the lower ones. These apical wear facets are planar elliptical surfaces oriented at approximately 45° to the long axis of the tooth. Unlike all other wear patterns observed in the Sauropoda, the wear facets on the teeth of *Diplodocus* are found on the labial sides of both upper and lower teeth (Barrett and Upchurch 1994; contra Calvo 1994a). The base of each facet (furthest from the apex) has a 'step' between the enamel and dentine (see above), whereas the top of the facet (near the tip of the crown) is 'flush.'

Fiorillo (1991) examined the microwear on the teeth of *Diplodocus*. Fine scratches were present, the majority being oriented 'perpendicular to the contact between the enamel and dentine' (Fiorillo 1991:24). No pits were found. Calvo (1994a), and our own SEM study, confirmed the presence of fine scratches extending labiolingually across the wear facets (Figure 4.5). In addition, there are pit-like areas (Figure 4.5), substantially larger than those typically formed by grit or phytoliths, which represent the results of extremely heavy wear.

Jaw joint and musculature The dorsal surface of the articular glenoid is strongly elongated anteroposteriorly, and there are no transverse ridges to hinder longitudinal movements of the quadrate over this surface (Barrett and Upchurch 1994). The elongate preorbital region, and the anteroventrally sloping quadrate, indicate that the jaw adductor muscles were oriented at a relatively low angle to the long axis of the skull. *Diplodocus* also has a long retroarticular process, which may reflect an increased role for M. depressor mandibulae in the jaw action; this muscle may have contributed to retraction of the lower jaw during a palinal movement and/or could have been involved in the evolution of a wider gape (see below).

Postcrania Diplodocids have increased the number of cervical vertebrae from 13 to 15 (*Diplodocus* and *Apatosaurus*) or 16 (*Barosaurus*). Individual vertebrae are elongate, especially in *Diplodocus* and *Barosaurus* (Upchurch 1995). As a result of these modifications, the necks of diplodocids are relatively long (Table 4.2), although they do not reach the extremes present in

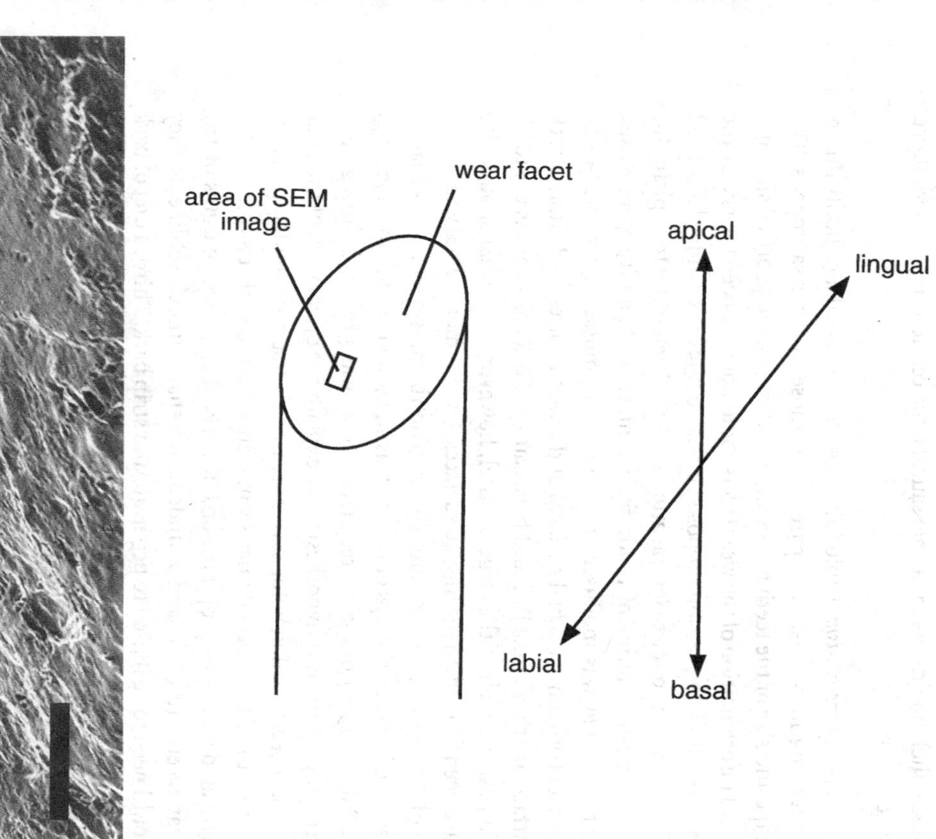

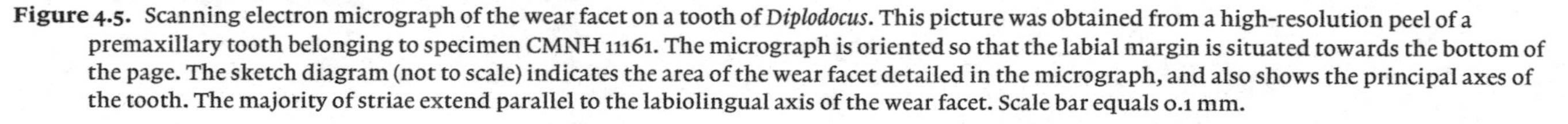

Figure 4.5. Scanning electron micrograph of the wear facet on a tooth of *Diplodocus*. This picture was obtained from a high-resolution peel of a premaxillary tooth belonging to specimen CMNH 11161. The micrograph is oriented so that the labial margin is situated towards the bottom of the page. The sketch diagram (not to scale) indicates the area of the wear facet detailed in the micrograph, and also shows the principal axes of the tooth. The majority of striae extend parallel to the labiolingual axis of the wear facet. Scale bar equals 0.1 mm.

some euhelopodids. As in dicraeosaurids, the cervical ribs are very short, especially in *Apatosaurus*.

Feeding mechanism Diplodocids clearly possessed a radically different feeding mechanism compared with those in other sauropods. The shape and size of the teeth makes them unsuitable for an orthal shearing bite. The orientation of the wear facets could not have been formed by the 'horizontal slicing' action proposed by Calvo (1994a), or indeed any kind of tooth–tooth contact. Several attempts have been made to explain this wear pattern in terms of tooth–food contact. Holland (1924) proposed that *Diplodocus* was molluscivorous, plucking mussels from shell-beds. Contact between the labial surfaces of the tooth apices, and either a rock surface or other mussels, could have caused the labial wear. As the teeth slid past the shell of the target mussel, however, some wear should also have been formed on the lingual surfaces. Furthermore, labial wear facets would have weakened the animal's grip on the mussel shell if the latter was held between the tips of the teeth as suggested by Holland (1924: fig. 4). Bakker (1986) proposed that labial wear was produced during bark stripping. While this hypothesis may explain the orientation of the wear facets, it seems unlikely for several reasons. First, the slender, procumbent teeth of *Diplodocus* do not seem well suited to withstand the forces generated by levering up portions of bark. Second, the margins of the wear facets are blunt and rounded, not sharp and chisel-like as they would need to be if used to penetrate and strip bark. Third, a diet of bark is very poor in nutrients (extant mammals only strip bark when other food sources become scarce).

Norman (1985) and Dodson (1990) have suggested that diplodocids fed by raking vegetation from branches. Barrett and Upchurch (1994) proposed a detailed model of unilateral branch stripping to explain labial wear and other features of the skull of *Diplodocus*. In unilateral branch stripping, one tooth row is used to strip foliage from a stem, while the other acts as a guide and stabilizer. Whether the upper or lower tooth row is used for stripping depends on the position of the animal's head relative to the stem, which, in turn, may reflect browse height (Barrett and Upchurch 1994: fig. 5). Thus the labial wear on the upper and lower tooth rows would be formed independently.

Unilateral branch stripping explains several features of the skull of *Diplodocus*. For example, given that the microwear was formed by tooth-to-food contact, the labiolingual orientation of the microscopic scratches indicates that the abrasive material moved parallel to the surfaces of the wear facets, in a direction lying within a parasagittal plane. The maxillary and dentary 'pseudo-cheek' would have prevented stripped foliage from escaping through the sides of the mouth during branch stripping. The elongate preorbital region would have allowed a longer portion of stem to be stripped in a single action. Palinal motion of the lower jaws (not proal, *contra* Barrett and Upchurch 1994) may have played two important roles: an increase in gape (facilitating placement of a foliage-bearing stem in the mouth) and fine adjustments of the relative positions of the tooth rows in order to produce a smooth stripping action.

The unilateral branch stripping model has implications for our understanding of feeding in *Diplodocus*. The orientation of the wear on the upper teeth suggests that stems were often raked at a steep upward angle relative to the long axis of the skull (Barrett and Upchurch 1994: fig. 5A). This angle, coupled with the greater robustness of the upper teeth and their heavier wear, suggests that high-browsing played an important role in food-gathering. The wear on the lower tooth row indicates that some food was acquired at lower browse heights (Barrett and Upchurch 1994). The elaborate nature of the proposed feeding mechanism for *Diplodocus* also implies more precise, and perhaps selective, food-gathering.

Diplodocids are often considered to have been high-browsers (Bakker 1971, 1986). The elongate neck would have allowed them to browse at maximum heights of 10 m (*Apatosaurus* and *Diplodocus*) or 12 m (*Barosaurus*) (Table 4.2). The shortened forelimbs and long sacral neural spines may have enabled these forms to rear up into a bipedal stance (Bakker 1986), increasing maximum browse height by as much as 2 m. The greater lateral flexibility of the neck, coupled with shortened forelimbs, could be interpreted as adaptations to food-gathering at lower levels. The high- and low-browsing models of diplodocid feeding are not mutually exclusive; indeed, such variation in vertical feeding range is consistent with unilateral branch stripping and increased food selectivity.

Implications for sauropod paleoecology

Many faunas, particularly in the Late Jurassic and Early Cretaceous, contained several sympatric sauropod taxa (Dodson 1990; Weishampel 1990). It is difficult to explain such diversity if sauropods are regarded as uniform high-browsers with simple jaw actions (Dodson 1990) because we might expect some ecological partitioning between similar organisms sharing the same environment. The data presented above indicate that the feeding mechanisms of sauropods were not as uniform as previously suggested. Possible niche separation between sauropods is illustrated in the following 'case study.'

To date, fourteen sauropod genera have been described from the Upper Jurassic Morrison Formation of western North America (McIntosh 1990), although no more than six have been recovered from any single locality (Dodson *et al.* 1980). Several localities include representatives of the 'Cetiosauridae' (*Haplocanthosaurus*), Brachiosauridae (*Brachiosaurus*), Camarasauridae (*Camarasaurus*) and Diplodocidae (*Apatosaurus, Barosaurus,* and *Diplodocus*). Each of the major sauropod groups in the Morrison Formation would have possessed a distinctive jaw mechanism. The camarasaurids, brachiosaurids, and probably the cetiosaurids (assuming the unknown skull of *Haplocanthosaurus* resembled that of *Patagosaurus*), had robust skulls and 'heavy,' spatulate dentitions suitable for feeding on coarse vegetation. Subtle variation in jaw action, such as the raking/tearing and precise cutting methods for foliage severing in camarasaurids and brachiosaurids, respectively, could be related to differences in preferred 'target' plants. Diplodocids possessed a highly unusual food-gathering system based on unilateral branch stripping. The relatively small head, narrow mouth, delicate tooth-comb and absence of oral processing imply that diplodocids fed on smaller and less coarse food items; they may have deliberately selected the more nutritious portions of plants, such as new shoots. This greater selectivity is consistent with the hypothesis that diplodocids fed over a wider vertical range than other sauropod groups. (The more nutritious parts of plants are scarcer and more widely distributed.)

Estimates of 'maximum' browse height provide some additional evidence for ecological partitioning of sauropods, although there may have been considerable overlap between genera. It seems likely that high-browsers would have often fed at levels below their maximum browse heights, and juveniles may have had to compete for resources at the same level as adult non-sauropods (such as ornithopods and stegosaurs). Even the adults of medium-sized sauropods may have browsed at similar heights as subadults of ultra-high browsers. Nevertheless, there does seem to have been some separation of sauropods on the basis of browse height. Brachiosaurids, and possibly *Barosaurus*, could have obtained food up to 14 m above ground level; camarasaurids and the remaining diplodocids were restricted to maximum browse heights of 10 m or less; and *Haplocanthosaurus* may not have fed above 5–6 m. The extent to which sauropods overlapped or were separated by browse height depends on a number of factors which are difficult to gauge: the relative costs and benefits of neck raising and lowering, and of bipedalism; the extent of overlap in preferred food items; the spatial distribution of preferred food items.

Several other dinosaur faunas, such as those in the Lower Shaximiao Formation (Middle Jurassic) of Sichuan (China) or Tendaguru beds (Late Jurassic) of Tanzania, contain three or more sympatric sauropod genera (Weishampel 1990). In each case, some ecological partitioning on the basis of feeding mechanism seems possible.

Phylogenetic patterns in sauropod feeding mechanisms

Ideally, we would like to understand the selective pressures and constraints that influenced sauropod evolution. Since food-gathering is such an important component in survival, it seems probable that patterns in feeding-mechanism evolution will reflect significant events in the history of the Sauropoda. In order to identify such patterns, relevant character-states have been mapped onto a simplified version of the cladogram presented earlier (Figure 4.6). Missing data inevitably produce some ambiguity in character-state distribution; the pattern discussed below is based on the delayed transformation optimization (DELTRAN) in PAUP 3.1.1 (Swofford 1993), which tends to favor convergence over reversal when ambiguity occurs. Thus the proposed sequence of apomorphy acquisition, and the identification of convergence versus reversal, should be treated as provisional.

Node A (Sauropoda) is characterized by an apomorphic lengthening of the forelimbs. This derived state is retained by all later sauropod groups, with partial reversals in *Shunosaurus* and the dicraeosaurid–diplodocid clade. *Brachiosaurus* has elongated forelimbs beyond the derived condition that unites all sauropods. Relative forelimb length may be affected by many factors, but the current distribution of this character within the sauropods suggests some correlation with browse height.

Several derived states combine to form the derived character-state known as 'spatulate teeth,' which occurs at node B. These apomorphic features include the development of a relatively larger and more mesiodistally expanded crown, the presence of one or two grooves on the labial surface, and the formation of a lingual concavity. These features persist throughout much of subsequent sauropod evolution but are lost at least

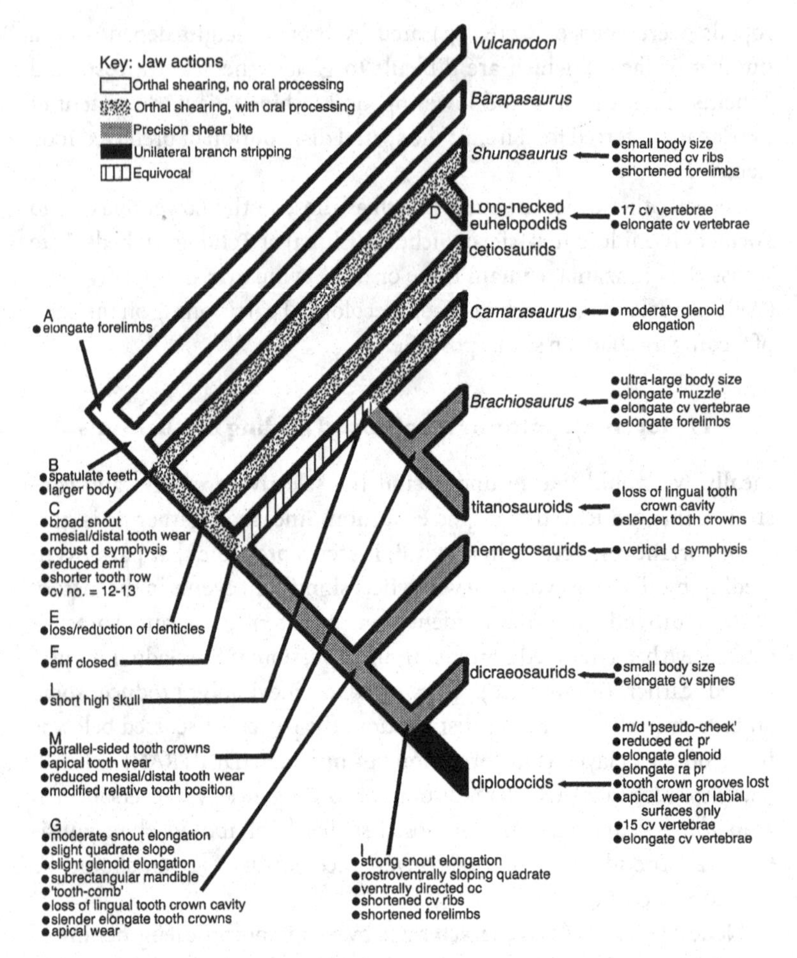

Figure 4.6. Feeding mechanism characters mapped onto a simplified sauropod cladogram (see text for details). The patterns within the branches of the cladogram show the distribution of different types of jaw action (see key). Abbreviations: cv, cervical; d, dentary; ect, ectopterygoid; emf, external mandibular fenestra; m, maxilla; oc, occipital condyle; pr, process; ra, retroarticular. For convenience, 'cetiosaurids' are represented as a single terminal taxon.

twice independently (see below). Most of these derived states can be interpreted as part of an adaptation to cropping coarse vegetation, although the functional significance of the lingual concavity is not understood. Node B may also be characterized by an increase in body size from 10 m to 15 m or more, representing a shift in sizes beyond that achieved by the largest prosauropods. Body size remains variable within the Sauropoda, with independent reductions in *Shunosaurus*, dicraeosaurids, and some titanosauroids, and independent occurrences of gigantism in *Brachiosaurus*, *Seismosaurus* and the titanosauroid *Argentinosaurus*. Changes in sauropod body size probably reflect a number of selective pressures, including those affecting food acquisition (such as browse height) and food processing (Farlow 1987).

Several important apomorphies are currently clustered at node C (Eusauropoda). This distribution may be misleading, however, because adequate vulcanodontid material is not available. It is likely, for example, that the acquisition of the robust mandibular symphysis and the increase in cervical number from 10 to 12 occurred below node C. Other apomorphic states, including the presence of mesial and distal tooth wear, seem to be genuinely absent in vulcanodontids. The derived character-states diagnosing the Eusauropoda are either retained or further elaborated in later sauropods; reversals to the original plesiomorphic conditions have not been identified.

Within the Euhelopodidae, derived states reflect specializations for different browsing heights. *Shunosaurus* developed a more flexible neck, shortened forelimbs, and smaller body size, as part of a low-browsing strategy. Other euhelopodids seem to have invaded a high-browsing niche: extremely long necks are achieved through the acquisition of an additional four cervical vertebrae and elongation of each cervical vertebra. Particularly elongate necks occurred independently two or perhaps three times elsewhere in the Sauropoda. Evidence in support of such convergence is provided not only by phylogenetic distribution, but also by the different methods used to achieve neck elongation: all groups exploit elongation of individual cervical vertebrae, but *Brachiosaurus* retains the plesiomorphic cervical number (12 or 13), euhelopodids 'create' new cervicals, and, in diplodocids, two dorsal vertebrae were integrated into the cervical series (Upchurch 1994). It is likely that titanosauroid necks also varied considerably in relative proportions, but the fragmentary preservation of known specimens currently obscures most of the useful details. There seems to have been repeated selective pressures favoring the development of long necks during sauropod evolution, perhaps reflecting competition between high-browsers or an 'arms race' with target food plants.

The only potential apomorphy at node E (cetiosaurids + neosauropods) is the loss of denticles on tooth crowns. The distribution of this character-state is complex: an independent loss of serrations occurs in *Euhelopus*, whereas small denticles are variably present in *Brachiosaurus*. The coarse serrations present in early sauropod teeth, as in prosauropods and ornithischians, may have facilitated the slicing/cutting of plant matter as tooth crowns slid past each other. Why cetiosaurids and neosauropods should have lost tooth serrations is not understood at present.

The external mandibular fenestra is reduced in eusauropods and finally obliterated in neosauropods (node F). This probably reflects a rearrangement of the adductor jaw musculature relative to vulcanodontids and prosauropods, but its functional significance is not clear.

The current lack of postcranial remains for nemegtosaurids means that only cranial apomorphies can cluster at node G (Diplodocoidea). Several apomorphies (Figure 4.6) mark a major shift in feeding mechanism relative to the cropping action present in other eusauropods. In particular, the acquisition of small 'peg'-like teeth arranged in a tooth-comb indicates the exploitation of a more delicate food source. Alterations to the shape of the mandible and the relative positions of teeth on opposing jaws would facilitate a more precise cutting action during feeding. This 'precise shear bite' is retained in dicraeosaurids, and an analogous system is acquired independently in Titanosauriformes. Although delayed transformation favors this 'convergence hypothesis,' it is equally parsimonious to suggest that the precision shear bite is a synapomorphy uniting the Neosauropoda, which underwent subsequent reversal in *Camarasaurus*. A similar ambiguity arises with respect to the elongation of the jaw joint and the utilization of propalinal jaw motion. Either slight propaliny was acquired independently in *Camarasaurus* and diplodocoids, or this derived character-state unites all neosauropods and is reversed in Titanosauriformes. Only the discovery of skull material belonging to basal camarasauromorphs and diplodocoids will resolve these issues.

Further modifications in the diplodocoid skull occur at node I. In particular, the role of propalinal jaw movements would have been enhanced by further elongation of the snout and the pronounced anteroventral orientation of the quadrate. Various adaptations for low-browsing were acquired, including shortened forelimbs and greater lateral flexibility of the neck. This implies that low-browsing was retained (with further modifications) by dicraeosaurids, whereas diplodocids superimposed a reinvented high-browsing strategy in order to achieve a wide vertical feeding range.

From the phylogenetic perspective, many components of the diplodocid feeding mechanism were acquired prior to unilateral branch stripping. For example, the elongate snout, anteroventrally oriented quadrate, and elongate articular glenoid, are also present in dicraeosaurids and are incipiently developed in nemegtosaurids. Although these features probably played an important role in unilateral branch stripping, they represent co-opted structures which were originally acquired as part of a precision shear jaw action. The slender 'pencil-like' teeth and maxillary-dentary 'pseudo-cheek,' however, may represent apomorphies specifically related to branch stripping.

The Camarasauromorpha (node L) are characterized by the development of a short, high skull. In basal forms, such as *Camarasaurus*, this modification may relate to a more vertical orientation of the adductor musculature, reflecting a powerful orthal shearing action. In the titanosauriform clade, however, there is a 'trend' towards the independent development of a more precise shear bite. This system included many of the apomorphies acquired by diplodocoids, but differs by the absence of a tooth-comb and propaliny. Some of the similarities between the diplodocoid and titanosauriform feeding mechanisms are the result of inevitable correlations between characters. For example, a more precise contact between opposing tooth rows automatically results in a reduction in mesial and distal wear and an increase in apical wear (Upchurch 1999). Why these two neosauropod clades should converge in some aspects of their feeding mechanics and diverge in others is not entirely clear. One possibility is that the similarities reflect a more selective diet, while the dissimilarities relate to preferences for different plants or parts of plants.

Current evidence implies that feeding mechanisms within the Titanosauroidea were themselves rather variable. For example, there is some indication that the skull of 'advanced' titanosaurs (e.g., *Saltasaurus*) had independently acquired some 'diplodocoid' features (Salgado and Calvo 1997; Upchurch 1999). The extent of such convergence, and patterns in apomorphy distribution, however, are obscured by the absence of both well-preserved titanosauroid cranial material and a well-supported phylogeny for this group.

The evolution of feeding mechanisms within the Sauropoda displays a complex pattern of convergence and reversal. Basic components of the jaw action (e.g., tooth shape and position, orientation of adductor jaw muscles, structure of jaw joint) seem relatively stable and long-lived. Most of the homoplasy in sauropod jaw mechanisms occurs as a result of the independent acquisition of precision shear bites in titanosauriforms and diplodocoids. In contrast, postcranial characters, such as relative neck and forelimb lengths, appear to be more variable during the course of sauropod evolution. There is at least a partial correlation between browse height and certain postcranial characters: elongate but inflexible necks, long forelimbs, and large body size are linked to high-browsing whereas more flexible necks, shortened forelimbs, and smaller body size are related to low-browsing. The reasons underlying these different levels of homoplasy are not well understood. This pattern could be an artefact created by the relatively poor preservation of many sauropod skulls; it may also represent a genuine phenomenon reflecting different selection pressures on skulls and postcrania.

The distribution of apomorphies on the sauropod cladogram, if taken at face value, implies three major phases in evolution of feeding mechanisms in sauropod dinosaurs. The first phase, occurring at nodes A–C, involved many modifications that would facilitate the cropping of large amounts of low-quality, coarse vegetation. The other two phases of 'rapid' evolution occur independently in the Diplodocoidea and Titanosauriformes; these apomorphies seem to reverse previous 'trends' insofar as they reflect more selective feeding on less coarse material. This interpretation has a number of important implications: (1) the initial radiation of sauropods in the Late Triassic and Early Jurassic may be related not only to the exploitation of a high-browsing niche, but also to the ability of these animals to process coarse, nutrient-poor vegetation; (2) most of the important innovations in sauropod feeding-mechanism evolution were acquired at a relatively early stage (i.e., by the time the major lineages had diverged in the Middle or early Late Jurassic); and (3) the 'decline' of 'broad-toothed' sauropods, during the Cretaceous, may have been triggered by some selection pressure(s) against forms that relied on large quantities of low-quality food. Thus fluctuations in sauropod diversity may have been strongly influenced by the evolution of feeding mechanisms. Such patterns should be treated with caution because of our current ignorance of much of sauropod evolution. For example, pattern (2) would be contradicted by the discovery that titanosauroids developed a wide variety of different feeding mechanisms during the Cretaceous. Nevertheless, current evidence suggests that feeding mechanisms played a central role in the origin and subsequent evolutionary history of sauropods.

Conclusions and prospectus

Without well-preserved, definite gastrointestinal contents, it will remain difficult to determine which plants, or parts of plants, formed the diets of different sauropod taxa. Nevertheless, the data presented above place limits on the probable food-gathering strategies employed by sauropods. Current conclusions include: (1) euhelopodids, cetiosaurids, and camarasaurids, used more oral processing than vulcanodontids and prosauropods; (2) a precise shearing bite was developed by brachiosaurids, titanosauroids, nemegtosaurids and dicraeosaurids; (3) diplodocids were specialized for branch stripping and may have been more selective feeders; (4) sauropods were probably specialized to exploit resources at different heights above ground level. Whatever interpretation we place on the data presented above, it seems clear that sauropod feeding mechanisms were more varied and complex than many previous studies have assumed. This diversity offers some insights into why many Jurassic and Cretaceous terrestrial faunas were able to support several sympatric sauropod taxa.

A variety of approaches could be used to expand this study. In particular, the tooth microwear in euhelopodids, brachiosaurids, and nemegtosaurids require attention. Furthermore, variation in macro- or microscopic wear patterns has not been examined systematically in teeth from different positions in the jaw or through ontogeny. A more rigorous engineering approach to the study of forces in the sauropod skull and postcranial skeleton should yield insights into the relationship between structure and function. There still exists no detailed understanding of the biomechanical significance of many of the cranial and vertebral characters that separate sauropods at the familial or generic level. By presenting our interpretations of sauropod feeding mechanisms here, we hope to encourage further investigation of these spectacular terrestrial herbivores.

Acknowledgments

This research was supported by The Leverhulme Trust, Bristol University (PU), a NERC studentship (GT4/93/128/G) and Trinity College, Cambridge (PMB). Grants from The Royal Society and The Dinosaur Society enabled us to collect data on Chinese sauropods. We are very grateful to the following for allowing access to the material in their care: Angela Milner and Sandra Chapman (BMNH), John Martin (LCM), Wolf-Dieter Heinrich and the late Hermann Jaeger (HMN), Mary Dawson, David Berman, Luo Zhexi and Norman Wuerthele (CMNH), Mark Norell (AMNH), Michael Brett-Surman (USNM), Cathy Forster (State University of New York, Stony Brook), Philip Powell (OUMZ), He Xin-Lu (CIT), Dong Zhiming (IVPP), Ouyang Hui (ZDM), Jose Bonaparte (MACN), and John

Peel (PMU). Lorraine Cornish (BMNH) gave expert advice on tooth casting. Ian Marshall (Department of Earth Sciences, Cambridge) and Bill Lee (Department of Zoology, Cambridge) greatly assisted our SEM studies. This paper was greatly improved as a result of helpful reviews by Peter Dodson and Jeff Wilson. We are extremely grateful to Xu Xing (IVPP) for his assistance during our research in China. Finally, we thank Hans-Dieter Sues for inviting us to contribute to this volume.

References

Alexander, R. M. (1985). Mechanics of posture and gait in some large dinosaurs. *Zool. J. Linn. Soc. London* 83:1–25.

Bakker, R. T. (1971). The ecology of the brontosaurs. *Nature* 229:172–174.

Bakker, R. T. (1986). *The Dinosaur Heresies*. New York: William Morrow.

Barrett, P. M. (1998). *Herbivory in the Non-Avian Dinosauria*. Unpublished Ph.D. Dissertation, University of Cambridge.

Barrett, P. M., and Upchurch, P. (1994). Feeding mechanisms of *Diplodocus*. *GAIA* 10:195–204.

Barrett, P. M., and Upchurch, P. (1995). Sauropod feeding mechanisms: their bearing on palaeoecology. In *Sixth Symposium on Mesozoic Terrestrial Ecosystems and Biotas, Short Papers*, eds. A. Sun and Y. Wang, pp. 107–110. Beijing: China Ocean Press.

Berman, D. S, and McIntosh, J. S. (1978). Skull and relationships of the Upper Jurassic sauropod *Apatosaurus* (Reptilia, Saurischia). *Bull. Carnegie Mus. Nat. Hist.* 8:1–35.

Bonaparte, J. F. (1986). Les Dinosaures (Carnosaures, Allosauridés, Sauropodes, Cétiosauridés) du Jurassique moyen de Cerro Condor (Chubut, Argentine). *Ann. Paléont.* 72:325–386.

Borsuk-Bialynicka, M. (1977). A new camarasaurid *Opisthocoelicaudia* gen. n. sp. n. from the Upper Cretaceous of Mongolia. *Palaeont. Pol.* 37:5–64.

Calvo, J. O. (1994a). Jaw mechanics in sauropod dinosaurs. *GAIA* 10:183–194.

Calvo, J. O. (1994b). Gastroliths in sauropod dinosaurs. *GAIA* 10:205–208.

Calvo, J. O., and Salgado, L. (1995). *Rebbachisaurus tesonei* sp. nov. A new Sauropoda from the Albian-Cenomanian of Argentina; new evidence on the origin of the Diplodocidae. *GAIA* 11:13–33.

Carey, M. A., and Madsen, J. H. Jr. (1972). Some observations on the growth, function and differentiation of sauropod teeth from the Cleveland-Lloyd Quarry. *Proc. Utah Acad. Sci., Arts, Lett.* 49: 39–43.

Choy, D. S. J., and Altman, P. (1992). The cardiovascular system of *Barosaurus*: an educated guess. *The Lancet* 340:534–536.

Christiansen, P. (1996). The evidence for and implications of gastroliths in sauropods (Dinosauria, Sauropoda). *GAIA* 12:1–7.

Coombs, W. P. (1975). Sauropod habits and habitats. *Palaeogeogr. Palaeoclimat., Palaeoecol.* 17:1–33.

Cooper, M. R. (1984). A reassessment of *Vulcanodon karibaensis* Raath (Dinosauria: Saurischia) and the origin of the Sauropoda. *Palaeontol. Afr.* 25:203–231.

Cope, E. D. (1877a). On *Amphicoelias*, a genus of saurian from the Dakota epoch of Colorado. *Proc. Am. Phil. Soc.* 17:242–246.

Cope, E. D. (1877b). On a gigantic saurian from the Dakota epoch of Colorado. *Paleontol. Bull.* 25:5–10.
Cope, E. D. (1878). A new species of *Amphicoelias. Am. Nat.* 12:563–565.
Dodson, P. (1990). Sauropod paleoecology. In *The Dinosauria,* ed. D. B. Weishampel, P. Dodson, and H. Osmólska, pp. 402–407. Berkeley: University of California Press.
Dodson, P., Behrensmeyer, A. K., Bakker, R. T., and McIntosh, J. S. (1980). Taphonomy and paleoecology of the Upper Jurassic Morrison Formation. *Paleobiology* 6:208–232.
Dong, Z. (1992). *Dinosaurian Faunas of China.* Berlin: Springer-Verlag.
Dong, Z., Zhou, S., and Zhang, Y. (1983). [The dinosaurian remains from Sichuan Basin, China.] *Palaeont. Sin., n.s. C* 162 (23):1–145 (in Chinese with English summary).
Farlow, J. O. (1987). Speculations about the diet and digestive physiology of herbivorous dinosaurs. *Paleobiology* 13:60–72.
Fiorillo, A. R. (1991). Dental microwear on the teeth of *Camarasaurus* and *Diplodocus:* implications for sauropod paleoecology. In *Fifth Symposium on Mesozoic Terrestrial Ecosystems and Biotas, Extended Abstracts,* ed. Z. Kielan-Jaworowska, N. Heintz, and H. A. Nakren, pp. 23–24. *Contr. Paleont. Mus., Univ. Oslo* 364.
Fiorillo, A. R. (1996). Further comments on the patterns of microwear and resource partitioning in the Morrison Formation sauropods *Diplodocus* and *Camarasaurus. J. Vert. Paleont.* 16(suppl. to 3):33A.
Gillette, D. D., and Lockley, M. G. (1989). *Dinosaur Tracks and Traces.* Cambridge and New York: Cambridge University Press.
Gilmore, C. W. (1925). A nearly complete articulated skeleton of *Camarasaurus,* a saurischian dinosaur from The Dinosaur National Monument. *Mem. Carnegie Mus. Nat. Hist.* 10:347–384.
Gilmore, C. W. (1932). On a newly mounted skeleton of *Diplodocus* in the United States National Museum. *Proc. U.S. Natl. Mus.* 81:1–21.
Gilmore, C. W. (1936). Osteology of *Apatosaurus* with special reference to specimens in the Carnegie Museum. *Mem. Carnegie Mus. Nat. Hist.* 11:175–300.
Haas, G. (1963). A proposed reconstruction of the jaw musculature of *Diplodocus. Ann. Carnegie Mus. Nat. Hist.* 36:139–157.
Hatcher, J. B. (1901). *Diplodocus* Marsh, its osteology, taxonomy and probable habits, with a restoration of the skeleton. *Mem. Carnegie Mus. Nat. Hist.* 1:1–64.
Hay, O. P. (1908). On the habits and the pose of the sauropodous dinosaurs, especially *Diplodocus. Am. Nat.* 42:672–681.
He, X-L., Li, C., and Cai, K. J. (1988). [*The Middle Jurassic dinosaur fauna from Dashanpu, Zigong, Sichuan: sauropod dinosaurs (2)* Omeisaurus tianfuensis.] Chengdu: Sichuan Publishing House of Science and Technology (in Chinese).
Holland, W. J. (1924). The skull of *Diplodocus. Mem. Carnegie Mus. Nat. Hist.* 9:379–403.
Huene, F. von (1929). Los Saurisquios y Ornithisquios del Crétaceo Argentino. *Ann. Mus. La Plata* 3:1–196.
Huene, F. von, and Matley, C. A. (1933). The Cretaceous Saurischia and Ornithischia of the central provinces of India. *Mem. Geol. Surv. India* 26:1–74.
Hunt, A. P., Lockley, M. G., Lucas, S. G., and Meyer, C. A. (1994). The global sauropod fossil record. *GAIA* 10:261–279.
Jacobs, L. L., Winkler, D. A., Downs, W. R., and Gomani, E. M. (1993). New material of an Early Cretaceous sauropod dinosaur from Africa. *Palaeontology* 36:523–534.
Jain, S. L., Kutty, T. S., Roychowdhury, T., and Chatterjee, S. (1975). The sauropod

dinosaur from the Lower Jurassic Kota Formation of India. *Proc. R. Soc. Lond. B* 188:221–228.

Jain, S. L., Kutty, T. S., Roychowdhury, T., and Chatterjee, S. (1979). Some characteristics of *Barapasaurus tagorei*, a sauropod dinosaur from the Lower Jurassic of Deccan, India. *Proc. Fourth Intl. Gondwana Symp., Calcutta 1977*, 1:204–216.

Janensch, W. (1929). Die Wirbelsäule der Gattung *Dicraeosaurus*. *Palaeontographica, Suppl.* 7(1), 2(1):37–133.

Janensch, W. (1935–36). Die Schädel der Sauropoden *Brachiosaurus, Barosaurus* und *Dicraeosaurus* aus den Tendaguru-Schichten Deutsch-Ostafrikas. *Palaeontographica, Suppl.* 7(1), 2(2):147–298.

Janensch, W. (1936). Ein aufgestelltes Skelett von *Dicraeosaurus hansemanni*. *Palaeontographica, Suppl.* 7(1), 2(2):299–308.

Janensch, W. (1950). Die Wirbelsäule von *Brachiosaurus brancai*. *Palaeontographica, Suppl.* 7(1), 3(2):27–93.

Janensch, W. (1961). Die Gliedmaszen und Gliedmaszengürtel der Sauropoden der Tendaguru-Schichten. *Palaeontographica, Suppl.* 7(1), 3(4):177–235.

Kellner, A. W. A. (1996). Remarks on Brazilian dinosaurs. *Mem. Queensland Mus.* 39:611–626.

Kurzanov, S. M., and Bannikov, A. F. (1983). [A new sauropod from the Upper Cretaceous of Mongolia.] *Paleontol. Zh.* 1983:91–97 (in Russian).

Lull, R. S. (1919). The sauropod dinosaur *Barosaurus* Marsh. *Mem. Connecticut Acad. Art. Sci.* 6:1–42.

Luo, Z., and Wu, X.-C. (1995). Correlation of vertebrate assemblage of the Lower Lufeng Formation, Yunnan, China. In *Sixth Symposium on Mesozoic Terrestrial Ecosystems and Biotas, Short Papers*, ed. A. Sun and Y. Wang, pp. 83–88. Beijing: China Ocean Press.

Madsen, J. H., McIntosh, J. S., and Berman, D. S. (1995). Skull and atlas-axis complex of the Upper Jurassic sauropod *Camarasaurus* Cope (Reptilia: Saurischia). *Bull. Carnegie Mus. Nat. Hist.* 31:1–115.

Marsh, O. C. (1877a). Notice of a new gigantic dinosaur. *Am. J. Sci.* (3) 14:87–88.

Marsh, O. C. (1877b). Notice of some new dinosaurian reptiles from the Jurassic Formation. *Am. J. Sci.* (3) 14:514–516.

Marsh, O. C. (1878). Principal characters of American Jurassic dinosaurs. Part I. *Am. J. Sci.* (3) 16:411–416.

Marsh, O. C. (1879). Principal characters of American Jurassic dinosaurs. Part. II. *Am. J. Sci.* (3) 17:86–92.

Marsh, O. C. (1883). Principal characters of American Jurassic dinosaurs. Part VI. Restoration of *Brontosaurus*. *Am. J. Sci.* (3) 26:81–85.

Marsh, O. C. (1884). Principal characters of American Jurassic dinosaurs, Part VII. On the Diplodocidae, a new family of the Sauropoda. *Am. J. Sci.* (3) 27:161–167.

Martin, J. (1987). Mobility and feeding of *Cetiosaurus* (Saurischia: Sauropoda) – why the long neck? In *Fourth Symposium on Mesozoic Terrestrial Ecosystems, Short Papers*, ed. P. J. Currie and E. H. Koster, pp. 154–159. Drumheller: Tyrrell Museum of Palaeontology.

Mateer, N. J., and McIntosh, J. S. (1985). A new reconstruction of the skull of *Euhelopus zdanskyi* (Saurischia, Sauropoda). *Bull. Geol. Inst. Univ. Uppsala* 11:125–132.

McIntosh, J. S. (1990). Sauropoda. In *The Dinosauria*, ed. D. B. Weishampel, P. Dodson, and H. Osmólska, pp. 345–401. Berkeley: University of California Press.

McIntosh, J. S., and Berman, D. S. (1975). Description of the palate and lower jaw of the sauropod dinosaur *Diplodocus* (Reptilia, Saurischia) with remarks on the nature of the skull of *Apatosaurus*. *J. Paleont.* 49:187–199.

McIntosh, J. S., Miller, W. E., Stadtman, K. L., and Gillette, D. D. (1996). The osteology of *Camarasaurus lewisi* (Jensen, 1988). *Brigham Young Univ. Geol. Stud.* 41:73–115.

Norman, D. B. (1985). *The Illustrated Encyclopaedia of Dinosaurs*. London: Salamander Books Ltd.

Nowinski, A. (1971). *Nemegtosaurus mongoliensis* n. gen., n. sp. (Sauropoda) from the uppermost Cretaceous of Mongolia. *Palaeont. Pol.* 25:57–81.

Osborn, H. F., and Mook, C. C. (1921). *Camarasaurus, Amphicoelias* and other sauropods of Cope. *Mem. Am. Mus. Nat. Hist.* 3: 247–287.

Owen, R. (1841). A description of a portion of the skeleton of the *Cetiosaurus*, a gigantic extinct saurian reptile occurring in the oolitic formations of different portions of England. *Proc. Geol. Soc. London* 3, Pt. II, No. 80:457–462.

Owen, R. (1842). Report on British fossil reptiles. Pt. II. *Rep. Brit. Ass. Adv. Sci.* 11:60–204.

Paul, G. S. (1988). The brachiosaur giants of the Morrison and Tendaguru with a description of a new subgenus, *Giraffatitan*, and a comparison of the world's largest dinosaurs. *Hunteria* 2(3):1–14.

Phillips, J. (1871). *Geology of Oxford and the Valley of the Thames*. Oxford.

Powell, J. E. (1987). Morfológia del esqueleto axial del los dinosaurios titanosauridos (Saurischia, Sauropoda) del estado de Minas Gerais, Brasil. *An. 10 Congr. Brasileiro Paleont.* 155–171.

Powell, J. E. (1992). Osteologia de *Saltasaurus loricatus* (Sauropoda-Titanosauridae) del Cretácico Superior des Noroeste Argentino. In *Los Dinosaurios y su entorno biotico*, ed. J. L. Sanz, and A. D. Buscalioni, pp. 165–230. *Act. Seg. Curso Paleontol. en Cuenca, Instituto 'Juan de Valdes.'*

Russell, D. A., and Zheng, Z. (1994). A large mamenchisaurid from the Junggar Basin, Xinjiang, People's Republic of China. *Can. J. Earth Sci.* 30:2082–2095.

Salgado, L., and Bonaparte, J. F. (1991). Un nuevo sauropodo Dicraeosauridae, *Amargasaurus cazaui* gen. et. sp. nov., de la Formacion la Amarga, Neocomiano de la Provincia del Neuquen, Argentina. *Ameghiniana* 28:333–346.

Salgado, L., and Calvo, J. O. (1992). Cranial osteology of *Amargasaurus cazaui* Salgado and Bonaparte (Sauropoda, Dicraeosauridae) from the Neocomian of Patagonia. *Ameghiniana* 29:337–346.

Salgado, L., and Calvo, J. O. (1997). Evolution of titanosaurid sauropods. II: The cranial evidence. *Ameghiniana* 34:33–48.

Salgado, L., Coria, R. A., and Calvo, J. O. (1997). Evolution of titanosaurid sauropods. I: Phylogenetic analysis based on the postcranial evidence. *Ameghiniana* 34:3–32.

Sereno, P. C., Dutheil, D. B., Larochene, M., Larsson, H. C. E., Lyon, G. H., Magwene, P. M., Sidor, C. A., Varricchio, D. J., and Wilson, J. A. (1996). Predatory dinosaurs from the Sahara and Late Cretaceous faunal differentiation. *Science* 272:986–991.

Swofford, D. L. (1993). *PAUP: Phylogenetic Analysis Using Parsimony, Version 3.1.1.* Champaign, IL: Natural History Survey.

Upchurch, P. (1993). *The Anatomy, Phylogeny, and Systematics of the Sauropod Dinosaurs.* Unpublished Ph.D. Dissertation, University of Cambridge.

Upchurch, P. (1994). Sauropod phylogeny and palaeoecology. *GAIA* 10:249–260.

Upchurch, P. (1995). The evolutionary history of sauropod dinosaurs. *Phil. Trans. R. Soc. Lond. B* 349:365–390.

Upchurch, P. (1998). The phylogenetic relationships of sauropod dinosaurs. *Zool. J. Linn. Soc.* 124:43–103.
Upchurch, P. (1999). The phylogenetic relationships of the Nemegtosauridae. *J. Vert. Paleont.* 19(1):106–125.
Weishampel, D. B. (1990). Dinosaurian distribution. In *The Dinosauria*, ed. D. B. Weishampel, P. Dodson, and H. Osmólska, pp. 63–140. Berkeley: University of California Press.
White, T. E. (1958). The braincase of *Camarasaurus lentus* Marsh. *J. Paleont.* 32:477–494.
Wilson, J. A., and Sereno, P. C. (1994). Higher-level phylogeny of sauropod dinosaurs. *J. Vert. Paleont.* 14(suppl. to 3):52A.
Wiman, C. (1929). Die Kreide-Dinosaurier aus Shantung. *Paleontol. Sin. C* 6:1–67.
Yadagiri, P. (1988). A new sauropod *Kotasaurus yamanpalliensis* from Lower Jurassic Kota Formation of India. *Rec. Geol. Surv. India* 11:102–127.
Young, C. C., and Chao, H. C. (1972). [*Mamenchisaurus hochuanensis* sp. nov.] *Inst. Vert. Paleontol. Paleoanthropol. Monogr. (Series A)* 8:1–30 (in Chinese).
Zhang, Y. (1988). [*The Middle Jurassic dinosaur fauna from Dashanpu, Zigong, Sichuan: Sauropod Dinosaurs (1)* Shunosaurus.] Chengdu: Sichuan Publishing House of Science and Technology (in Chinese).
Zheng, Z. (1991). [Morphology of the braincase of *Shunosaurus*.] *Vert. PalAsiat.* 29:108–118 (in Chinese with English summary).

DAVID B. WEISHAMPEL AND CORALIA-MARIA JIANU

5

Plant-eaters and ghost lineages: dinosaurian herbivory revisited

Introduction

Biotic interactions, whether they are in the form of hosts and parasites, flowers and pollinators, or seeds and dispersers, are often considered the *sine qua non* of coevolution, a process that some have argued has an overarching influence on the structure of life on Earth (Futuyma and Slatkin 1983; Margulis and Fester 1991; Thompson 1994). Whether true or not, the identification of coevolutionary relationships can be rather difficult from a paleontological perspective, particularly when considering groups of extinct organisms well removed from those of the present day.

Several attempts have been made to understand plant–herbivore interactions during the Mesozoic, a time when plant-eaters interacted with a mixture of 'pteridophyte', 'gymnosperm', and angiosperm plants. Virtually all of these studies have begun by grouping co-occurring faunas and floras (e.g., Benton 1984; Hotton 1986; Tiffney 1986; Farlow 1987; Coe *et al.* 1987; Fleming and Lips 1991). Co-occurrence is either in terms of taxa that may have associated with one another or the features that reflect this plant–herbivore interaction (tooth shape, foliage type, etc.). In more expanded form, these groupings are then examined in light of their paleobiogeographic and temporal distributions to understand their coevolutionary patterns (e.g., Bakker 1978; Wing and Tiffney 1987).

The discovery of patterns of ecological interactions and/or coevolution between plants and herbivores during the Mesozoic is particularly important in view of the great radiation of angiosperms beginning in the latter half of the Early Cretaceous and extending through the end of that period (Doyle and Donoghue 1986; Crane 1989). In view of these changes in the plant realm, not only in diversity, but also in physiognomy and life histories (Crane 1987; Upchurch and Wolfe 1987), contemporary herbivores surely confronted new feeding opportunities and perhaps problems associated with their digestion. Consequently, it is important to evaluate changes in taxonomic diversity and feeding systems among contemporary primary consumers of these angiosperms and other plants.

An early attempt to identify such changes was made by Weishampel and Norman (1989); this study examined the evolutionary dynamics of herbivorous tetrapods across the Mesozoic. This work used species of a variety of monophyletic higher taxa, including Procolophonoidea, Synapsida (e.g., Dicynodontia, Tritylodontoidea, and Multituberculata), Rhynchosauria, Aetosauria (Stagonolepididae), and Dinosauria and, while each group was treated as monophyletic, there was no attempt to incorporate more detailed phylogenetic information beyond combining species into higher taxa. Inferences were made about the feeding behavior of members of these clades on the basis of body size, tooth structure, and jaw mechanics, among other anatomical features (see below).

From these kinds of data (taxon-based and function-based), Weishampel and Norman (1989) estimated speciation, extinction, and species-based net-profit-and-loss rates. These metrics were compared with similar dynamics among terrestrial plants as documented by Niklas, Tiffney and Knoll (1980, 1985), in order to identify the evolutionary dynamics among these herbivores and plants through the Mesozoic. For example, many ornithischians – the most taxonomically and functionally diverse group among all Mesozoic herbivorous tetrapods – appear to show a rise in species origination toward the end of the Early Cretaceous and another increase toward the end of the Late Cretaceous, a pattern which was interpreted as roughly mirroring the radiation of early angiosperms.

If this ornithischian pattern proves to be real, then it suggests that ornithischians may have had a close adaptive if not coevolutionary connection with emerging angiosperms. Expressed in another way, might angiosperms have been a kind of forcing factor in the trophic diversification of ornithischians? In order to answer this question, we need as complete as possible a census of ornithischians and other Mesozoic herbivores. The fossil record provides the ultimate basis for this census (i.e., raw species counts), but because taphonomic biases are unpredictable, it is unclear what these numbers may mean. One way to approach a more accurate census is to include diversity data that come not only from the

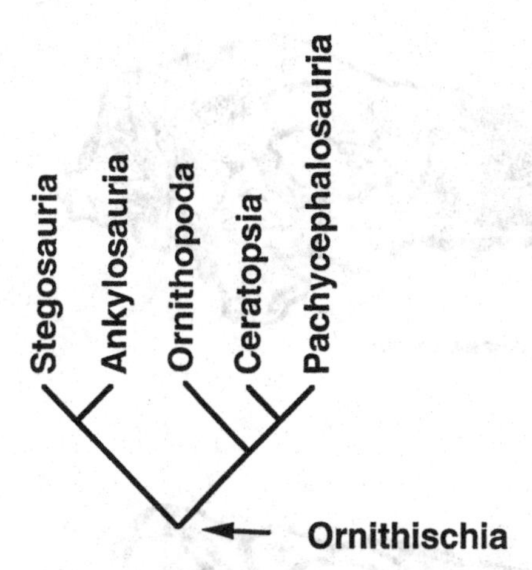

Figure 5.1. Phylogenetic relationships of the major clades within Ornithischia (redrawn from Sereno 1986).

actual census of Mesozoic herbivores and their feeding styles, but also from the identification of ghost lineages.

Taxa and feeding groups

In order to examine the relationship between taxa known directly from the fossil record and the ghost lineages inferred from them on estimates of diversity among Mesozoic herbivores, we have concentrated on the two major clades of dinosaurian herbivores, Ornithischia and Sauropodomorpha. These clades are well represented throughout the Mesozoic, species diversity is often very high, and phylogenetic analyses of most of the taxa within these clades are now emerging.

Five major clades are known within Ornithischia (Figure 5.1), with additional clusters of species intercalated among them. The first major ornithischian clade under consideration is Thyreophora, consisting of Ankylosauria and Stegosauria, with several individual stem species. The Ankylosauria (Figure 5.2a) consist of quadrupedal, heavily armored dinosaurs, which are known primarily from the Cretaceous (although from as early as the Middle Jurassic) of North America, Europe, Asia and Australia. Ranging in length from 5 m to 9 m, these herbivores browsed within 1–2 m off the ground and used the cutting edges of their scoop-shaped beak to crop foliage and perhaps fructifications to be delivered to the small, simple, triangular teeth that lined the jaws (Coombs and Maryanska 1990).

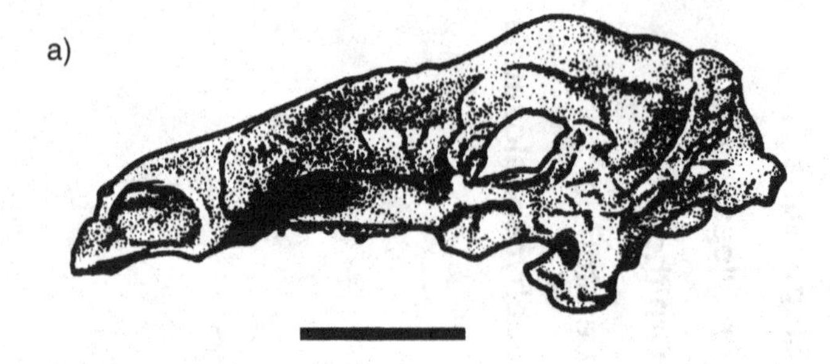

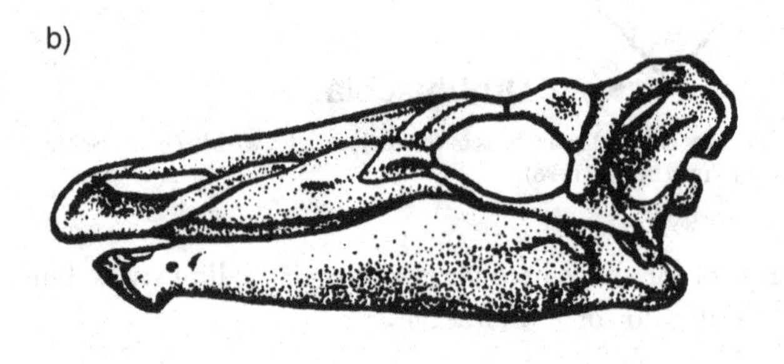

Figure 5.2. Thyreophora. a) Skull of the nodosaurid ankylosaur *Panoplosaurus mirus* (without lower jaw). b) Skull of the stegosaur *Stegosaurus stenops*. Scale bar equals 10 cm.

These teeth bear prominent denticles and are somewhat separated from each other; chewing appears to have taken place by simple jaw closure (i.e., an orthal power stroke). Weishampel and Norman (1989) identified this kind of dentition as indicative of puncture-crushing mastication. Tooth wear is often present, but not systematically developed. The oral cavity was bound by muscular cheeks (as in all ornithischians except *Lesothosaurus diagnosticus*), and there may have been a long flexible tongue. The enormous abdominal region suggests a large fermentation compartment.

The other major thyreophoran clade, Stegosauria (Figure 5.2b), is known primarily from the Middle Jurassic through the Early Cretaceous of Europe, Asia, North America and Africa. These heavily built, 3–9 m long quadrupeds bore a variety of spines and/or plates along the ridge of their back. Stegosaurs had relatively small but elongate heads with simple, spatulate, denticulate teeth with relatively large gaps between teeth. These teeth occasionally display obliquely inclined wear surfaces, suggesting at least some sort of orthal puncture crushing like that seen in

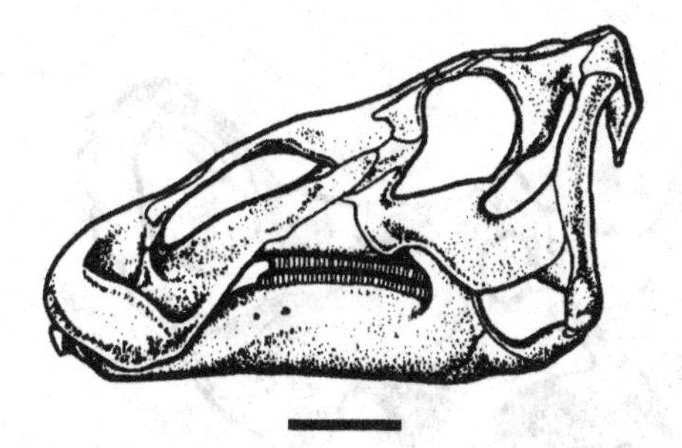

Figure 5.3. Ornithopoda. Skull of the hadrosaurid *Edmontosaurus regalis*. Scale bar equals 20 cm.

ankylosaurs (Galton 1985, 1990). Browsing was likely within a meter above the ground although it is possible that some stegosaurs may have been able to browse at higher (2–3 m) levels by assuming a tripodal posture.

Ornithopoda (Figure 5.3), the longest-lived ornithischian clade (Early Jurassic through the end of the Cretaceous), is known worldwide. Ranging in length from 1 m to 12 m, these typically bipedal herbivores had closely packed teeth with well-developed medium-angle shear surfaces that extended continuously along the length of the dentition. Consequently, ornithopods can be characterized as having a transverse grinding style of mastication (Weishampel 1984; Norman 1984; Weishampel and Witmer 1990; Sues and Norman 1990; Norman and Weishampel 1990; Weishampel and Horner 1990). Browsing appears to have been concentrated within 1–2 meters off the ground, although in the large iguanodontians (e.g., *Iguanodon*, most hadrosaurids) it may have extended up to 4 m.

Known nearly exclusively from the Cretaceous from the Northern Hemisphere, ceratopsians (Figure 5.4a) were mostly quadrupedal herbivores. These so-called horned dinosaurs ranged in length from 2 m to 9 m and browsed no more than 2 m above the ground. The front of the jaws was equipped with a hooked parrot-like beak and the cheek teeth were closely packed, worn into a continuous vertical chewing surface (Ostrom 1964; Dodson and Currie 1990). Only in species of *Psittacosaurus* is the occlusal surface worn at a lower angle (approximately 60°) and this basal ceratopsian is known to possess gastroliths (Sereno 1990).

Pachycephalosaurs (Figure 5.4b), like their marginocephalian sister-group Ceratopsia, are known only from the Cretaceous of the Northern Hemisphere. Ranging from 2 m to 5 m in length, these domed-headed

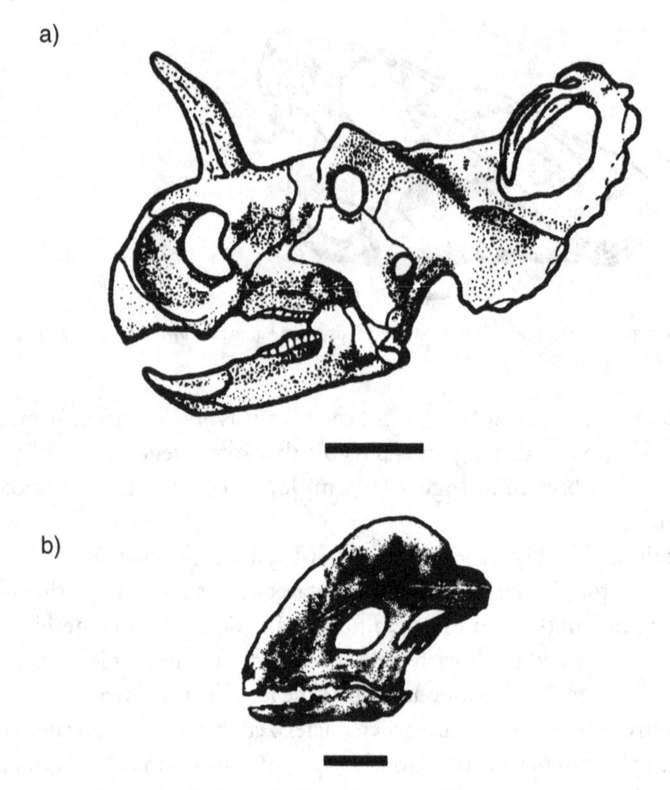

Figure 5.4. Marginocephalia. a) Skull of the ceratopsid ceratopsian *Centrosaurus apertus*. Scale bar equals 20 cm. b) Skull of the pachycephalosaur *Stegoceras validum*. Scale bar equals 5 cm.

ornithischians were foragers on low-growing plants (1 m); the small, triangular, denticulate teeth had relatively large gaps between them and an orthal power stroke (Maryanska 1990). Finally, the large abdominal region suggests the presence of a large fermentation chamber.

The other great clade of dinosaurian herbivores, Sauropodomorpha (Figure 5.5), is known worldwide from the Late Triassic through the end of the Cretaceous. The major clades of sauropodomorphs include prosauropods, euhelopodids, camarasaurids, brachiosaurids, titanosauroids, and diplodocoids, with other taxa interpolated among them (Upchurch 1995, 1998). Although new research suggests that many were clearly not restricted to plant-eating (Barrett, this volume), sauropodomorphs ranged in length from 2.5 m to perhaps as much as 30 m (McIntosh 1990; Dodson 1990a). The dentition consisted of relatively widely spaced,

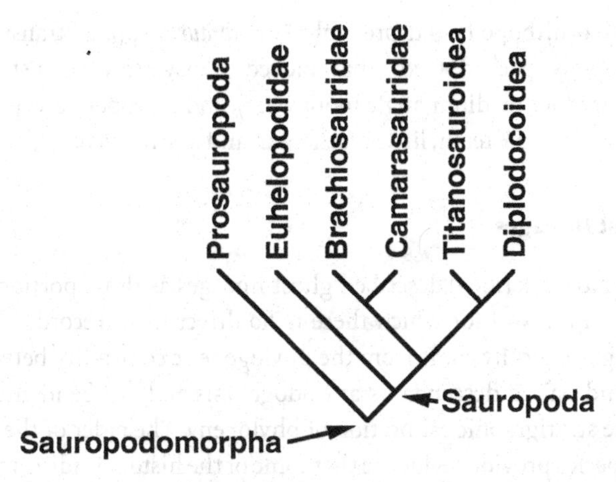

Figure 5.5. Phylogenetic relationships of the major clades within Sauropodomorpha (redrawn from Upchurch 1995, 1998).

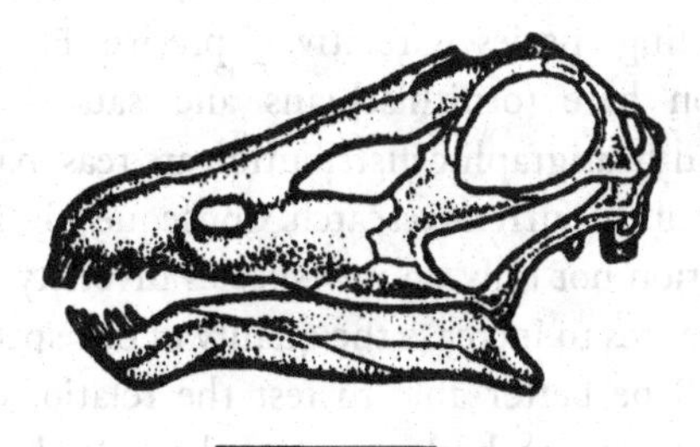

Figure 5.6. Sauropodomorpha. Skull of the diplodocid *Diplodocus longus*. Scale bar equals 20 cm.

simple peg-like or spatulate teeth often restricted to the front of the snout (Figure 5.6; Calvo 1994; Barrett and Upchurch 1994). Tooth wear is known in several sauropodomorphs (Fiorillo 1991) and indicates that foliage was stripped from branches but not extensively chewed. Mechanical breakdown of foliage instead appears to have been accomplished by gastroliths in a muscular gizzard. Browsing was at high levels (perhaps up to 15 m).

Thus we can characterize the various clades of ornithischians and sauropodomorphs in the following way (Weishampel and Norman 1989): stegosaurs, ankylosaurs, and pachycephalosaurs are envisioned as orthal pulpers (denticulate crowns, large gaps between teeth, and an orthal power stroke), ceratopsians excluding *Psittacosaurus* spp. as orthal slicers (closely packed teeth and a vertical power stroke that produced high-angle wear facets), ornithopods and probably *Psittacosaurus* spp. as transverse grinders (closely packed teeth that moved transversely against one another to produce medium-angle tooth wear), and sauropodomorphs as gut processors (simple teeth, little tooth wear, and gastroliths).

Ghost lineages

Norell and Novacek (1993) described ghost lineages as those portions of the history of a taxon for which there is no direct fossil record. These ghost lineages logically come from the phylogenetic continuity between ancestors and paired descendants at cladogenesis and can be identified through the stratigraphic calibration of phylogeny. The older of the two resulting species provides evidence that some of the history leading to the sister species is missing from the stratigraphic record.

Ghost lineages are most accurately identified when the phylogeny of the group of interest is well understood and the stratigraphic occurrence of the descendant sibling species is relatively precise. For the groups under consideration here (ornithischians and sauropodomorphs), both phylogeny and stratigraphic distribution are reasonably well known or at least the focus of current research. Consequently, it is possible to include information not only on raw species diversity but also on ghost lineage occurrences to increase the quality of our species census. In this way, we will be better able to test the relationship between the evolution of dinosaurian herbivory and the radiation of angiosperms.

Raw and ghost diversity patterns

Because they were the most diverse and abundant among Mesozoic herbivores and because their phylogeny is reasonably well understood, ornithischians and sauropodomorphs are the best subjects for analyzing possible adaptive and/or coevolutionary relationships with contemporary plants. The species-level cladogram for Ornithischia used in this study is based in part on Sereno (1986) and includes information from Sues and Galton (1987), Coombs and Maryanska (1990), Sereno and Dong (1992), Weishampel, Norman and Grigorescu (1993), Forster *et al.* (1993), Sampson (1995), Weishampel (1996), and Forster (1996). Resolution of species on this cladogram are generally unproblematic as most genera are monospecific. Where multispecific genera are encountered, they are positioned as unresolved sister-taxa, sometimes with a star-burst effect. The total number of species on this ornithischian cladogram is nearly 140, ranging in age from the Early Jurassic (Hettangian–?Sinemurian) through the end of the Cretaceous (late Maastrichtian; Weishampel 1990a). The species-level cladogram is available from the authors on request.

The species-level cladogram for Sauropodomorpha is more problematic, and we rely on Galton (1990) and Upchurch (1995, 1998) with interpolation of additional species from McIntosh (1990). As in ornithischians, single species are most commonly allotted to each sauropodomorph genus. The total number of sauropodomorph species on this cladogram is nearly 100, ranging in age from the Late Triassic (Norian) through the end of the Cretaceous (late Maastrichtian; Weishampel 1990a). The species-level sauropodomorph cladogram is available from the authors on request.

These phylogenetic relationships are then calibrated against the earliest occurrence of each species. Ghost lineages of both ornithischians and sauropodomorphs are rather scattered throughout each clade, but are an integral part of all higher taxa. Figure 5.7 provides a detail taken from Iguanodontia, one of the most diverse clades of Ornithopoda.

In order to assess the relationship of ghost lineages to estimates of ornithischian diversity, we sampled actual species and their ghost lineages at 2.5-million-year intervals from the end of the Cretaceous back through to the earliest occurrence of both major clades of dinosaurian herbivores. This 2.5-million-year interval was chosen in order to insure that all species as well as their ghost lineages will be sampled in view of Dodson's (1990b) calculation that dinosaurian species have a temporal range of approximately 5 million years.

The raw count of ornithischian species (Figure 5.8) reveals a modest diversity peak in the Early Jurassic, a second much larger series of peaks in the Late Jurassic, another peak at the end of the Early Cretaceous, and finally the largest peak of the series at the end of the Cretaceous, much as was identified by Weishampel and Norman (1989). Among sauropodomorphs, raw diversity levels are much more modest than for ornithischians. There is a peak in the Late Jurassic, another less substantial peak in the mid-Cretaceous, and another just prior to the end of the Cretaceous (Figure 5.9).

In contrast, when the phylogeny of these same species is calibrated against stratigraphy, the pattern of ornithischian ghost lineages indicates

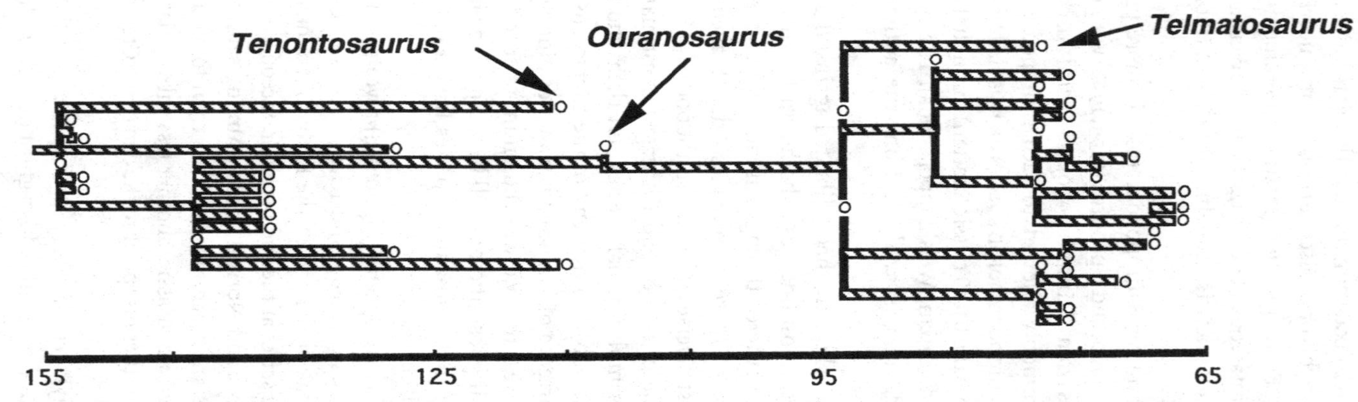

Figure 5.7. Ghost-lineage diagram for Iguanodontia. The circles represent the stratigraphic occurrences of iguanodontian species (*Tenontosaurus tilletti*, *Ouranosaurus nigeriensis*, and *Telmatosaurus transsylvanicus* are indicated). The vertical lines represent relationships among species, and the striped horizontal lines constitute ghost lineages.

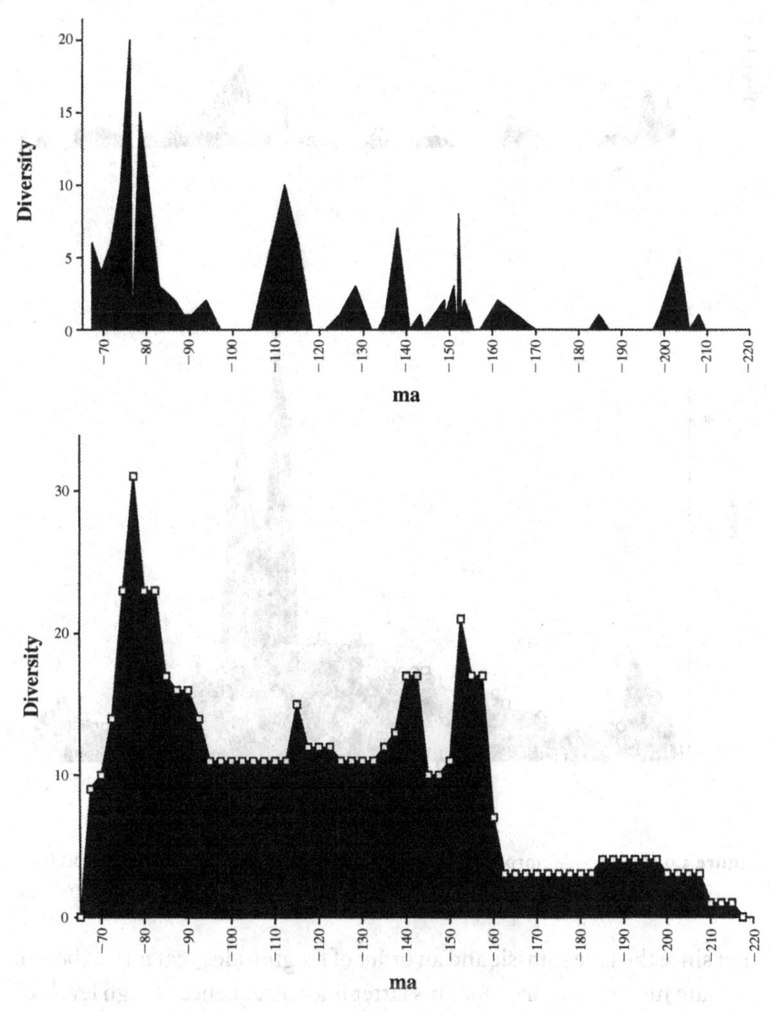

Figure 5.8. Ornithischian diversity through the Mesozoic, based on raw species data from the fossil record (above) and based on ghost-lineage data (below).

a more-or-less gradual rise in diversity from the Early to Late Jurassic (represented principally by stegosaurs), followed by an abrupt increase in the Late Jurassic (principally driven by stegosaurs and basal iguanodontians), followed by a reduction in diversity that levels across the mid-Cretaceous (broadly distributed among all ornithischian clades except stegosaurs), followed by the greatest diversity increase at the end of the Cretaceous, in which the summed species diversity is more than twice

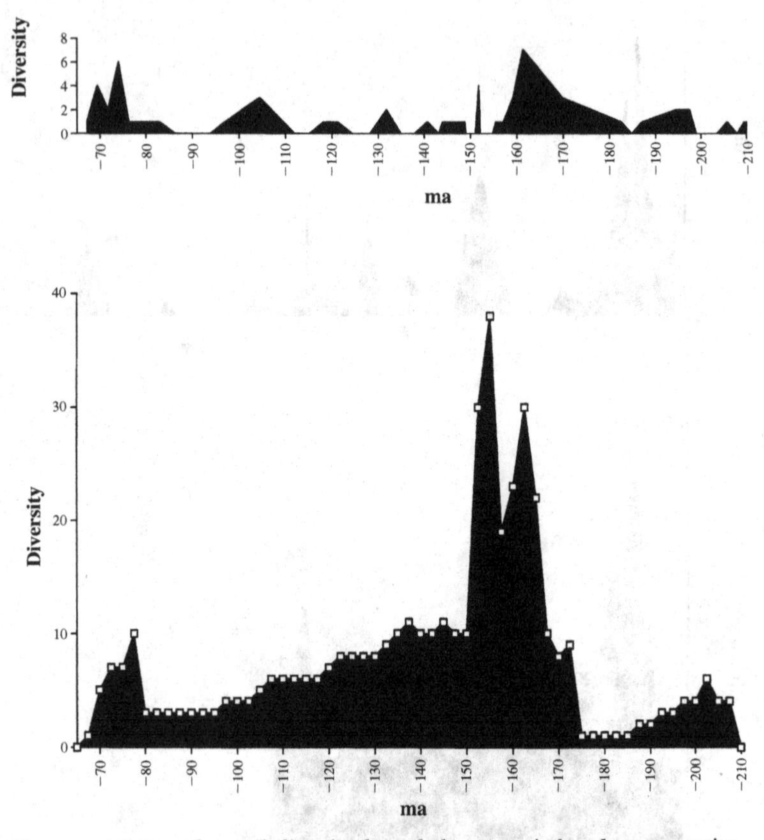

Figure 5.9. Sauropodomorph diversity through the Mesozoic, based on raw species data from the fossil record (above) and based on ghost-lineage data (below).

that since the Late Jurassic and an order of magnitude greater than before the Late Jurassic (Figure 5.8). This latter is a consequence of high levels of cladogenesis among hadrosaurid ornithopods and ceratopsid ceratopsians. For sauropodomorphs (Figure 5.9), there is a small peak at the end of the Triassic, another slightly higher peak at the beginning of the Jurassic (together based on cladogenesis of prosauropods), followed by a decline through most of the Middle Jurassic, then the greatest peak at the end of the Jurassic, during which time diversity jumped approximately sevenfold (driven by increased cladogenesis in brachiosaurids, camarasaurids, titanosauroids and diplodocoids). Thereafter, diversity levels gradually decline through the end of the Mesozoic, with a final abrupt decline from the Campanian through the Maastrichtian.

Feeding group diversity

However interesting these taxonomic data may be, it is really the diversity of feeding groups that is ultimately useful in assessing herbivore–plant interactions and/or coevolution. Consequently, we have taken the feeding styles that have been identified among known species and deduced 'ghost feeding styles' from their ghost lineages. As described previously, these feeding styles include orthal pulping (*Lesothosaurus*, thyreophorans, pachycephalosaurs), orthal slicing (nearly all ceratopsians), transverse grinding (ornithopods, *Psittacosaurus*), and gut processing (sauropodomorphs, *Psittacosaurus*). Using these categories, ornithischian and sauropodomorph feeding styles were mapped and optimized onto the cladogram for Ornithischia using the DELTRAN optimization option of PAUP (Swofford 1985) in order to produce the minimal resolution of feeding styles on the tree given the available data. Because of the strictures of DELTRAN and lack of appropriate information from the fossils themselves, we also have ghost lineages with unresolved feeding styles. Among others, this includes the feeding style of the ghost lineage leading from the base of Ornithischia to Neoceratopsia.

Sampling of optimized feeding styles was again at the same 2.5 million years as in the ghost lineage analyses described previously.

Ghost-lineage diversity of these feeding styles is indicated in Figure 5.10. From the end of the Late Triassic until the Late Jurassic, gut processors, orthal pulpers, and transverse grinders existed at relatively low diversity, although gut processors were at least twice as diverse as the other two feeding styles. All three groups show an abrupt diversity increase in the Late Jurassic, gut processors preceding orthal pulpers and transverse grinders by approximately 10 million years. Late Jurassic diversity of gut processors is twice that of both other groups combined. Thereafter, diversity levels decreased among all three groups, throughout the remainder of the Mesozoic in gut processors, and through the middle of the Early Cretaceous in orthal pulpers and transverse grinders. Orthal pulping then appears to have undergone a second wave of diversification in the Late Cretaceous. In contrast, transverse grinders have a modest peak at the end of the Early Cretaceous, followed by a decline through the first half of the Late Cretaceous and a final rise at the end of the Mesozoic. The last feeding style, orthal slicing, is restricted to the Late Cretaceous. This short-lived mode of oral processing goes from a diversity of zero to eight species to zero again in the course of 25 million years.

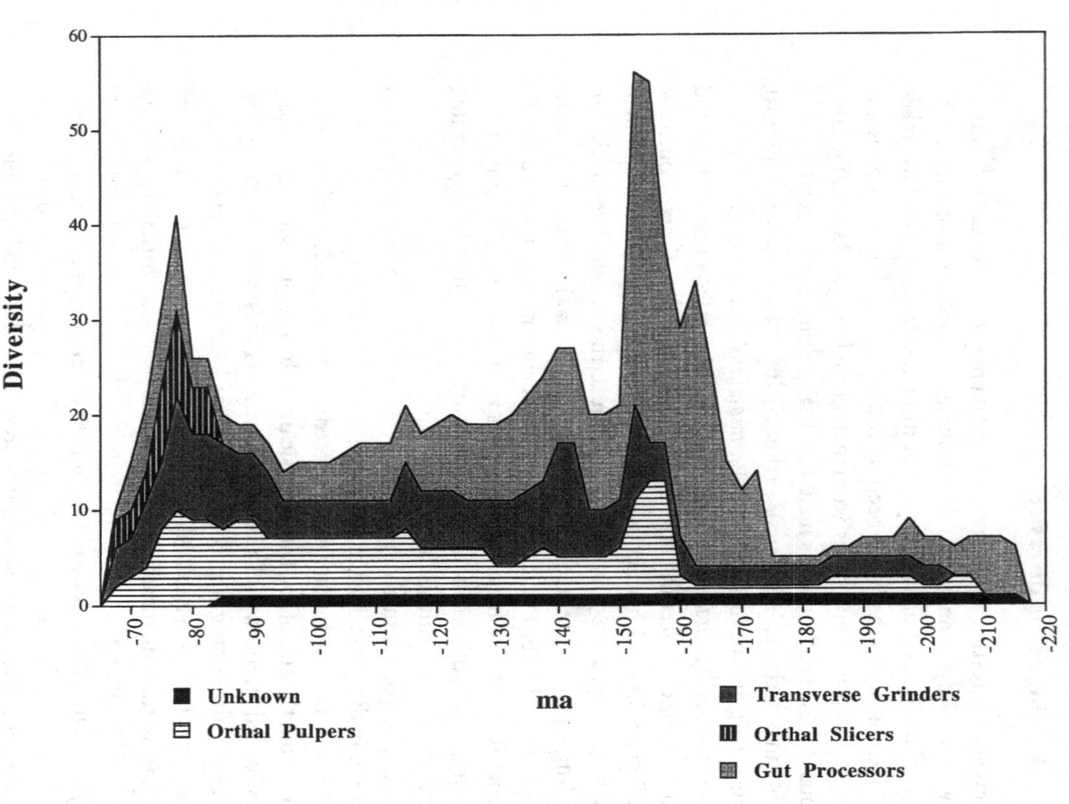

Figure 5.10. Feeding style diversity among herbivorous dinosaurs through the Mesozoic, based on ghost-lineage data.

In order to reduce the effects of different sample sizes and the patchiness of certain sampling intervals, we have also transformed these data into percentage contributions of feeding styles across these same intervals (Figure 5.11). The Late Triassic began solely with gut processors, which gradually gave way to orthal pulpers and transverse grinders through the Early Jurassic. From the Middle through the Late Jurassic, gut processors reverse this pattern, with a concomitant percentage decrease in orthal pulpers and transverse grinders. Transverse grinders maintain a fairly stable diversity percentage (approximately 20–30%) through the Late Cretaceous. At the same time, orthal pulpers show a modest increase in diversity in the mid-Cretaceous. In contrast, gut processors steadily decline from the Early Cretaceous onward. Finally, the rise in orthal slicers in the Late Cretaceous contributes a large percent diversity increase, apparently involving reciprocal percent declines in orthal pulpers and gut processors but not transverse grinders.

Discussion

The use of ghost lineage diversity, rather than raw species counts, provides a more accurate picture of minimal diversity levels at any time in the geological past. Due to the phylogenetic continuity of species within clades, it is possible to deduce minimal levels of species diversity – as well as feeding styles – beyond those species physically available to us from the fossil record. Without this additional ghost-lineage-based information, raw species counts and evolutionary metrics derived from them can provide only a partial picture of changes in herbivore diversity across the Mesozoic. For example, raw species counts from Ornithischia underrepresent minimal diversity levels overall by 75% and by as much as 95% for individual intervals (Figure 5.8). Even at their best, these raw counts do only as well as 50% representation when sample size is low and the clade is young.

Incorporation of information from ghost lineages affects not only relative diversity levels at any given moment, but, as a consequence, also affects overall patterns of change within a given clade. Norman and Weishampel (1989) indicated three separate intervals of diversification among transverse grinders (as represented by 'advanced ornithopods,' now referred to as euornithopods; Weishampel 1990b), with the middle peak arguably associated with the origin of angiosperms. Based on ghost-lineage diversity, however, this middle peak virtually disappears. In its

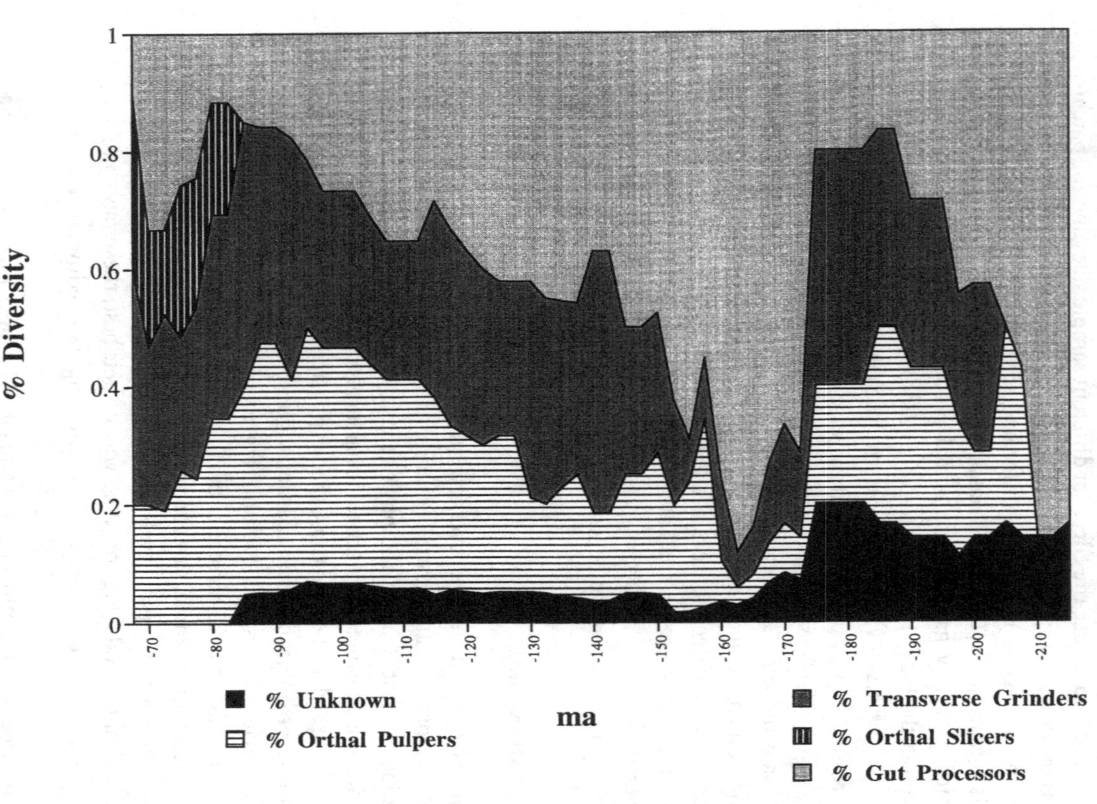

Figure 5.11. Feeding style diversity among herbivorous dinosaurs through the Mesozoic, expressed as percentages.

place, there are only modest fluctuations in diversity that intercede between the extreme Late Jurassic and Late Cretaceous peaks.

Given these revised metrics, what is the relationship of the Early Cretaceous origin of angiosperms to changes in feeding styles? Species diversity of both orthal pulpers and perhaps transverse grinders based on ghost lineages increases only modestly if at all through this time when angiosperms appear to have begun their initial radiation. This pattern exists in both untransformed and percentage analyses (cf. Figures 5.10–5.11). If we regard diversity patterns as a signal for important evolutionary events linking plants and herbivores, then fluctuations in ghost-lineage diversity are less than spectacular, given the major radiation among angiosperms. Instead, it is during the Late Cretaceous that diversity is radically altered, in greatest part due to the origin and diversification of orthal slicers, indicated in both ghost diversity counts and percentages, but also increases in diversity among transverse grinders and slightly less so among orthal pulpers, indicated in ghost diversity counts but not percentages.

Whatever coevolutionary interactions may have existed between flowering plants (angiosperms) and ornithischian dinosaurs, they are not apparent at the onset of the evolutionary radiation of angiosperms, as least as far as diversity patterns are concerned. Whether coevolutionary links existed, but lagged behind the Early Cretaceous diversification of flowering plants is equally unclear. One might imagine that the Late Cretaceous diversification of orthal pulpers, transverse grinders, and orthal slicers (based on untransformed ghost diversity) were reflected in changes in Late Cretaceous angiosperms, such as alterations in their physiognomy, as per shifts in stature, or in threshold effects such as floral dominance or shifting biogeographic realms (but see Wing, Hickey and Swisher 1993). Conversely, changes in food resources provided by other tracheophytes might explain these diversity patterns among contemporary herbivores. However, these are merely suggestions and at present nothing explicitly links Mesozoic plants and herbivores in a coevolutionary way beyond these untested scenarios.

What we have presented here is, in part, a revision of earlier work on the trophic diversity of herbivorous dinosaurs (Weishampel and Norman 1989), as well as an advocacy of the use of phylogeny to provide additional information on such diversity. Ornithischians and sauropodomorphs appear to have been little affected by the initial radiation of angiosperms, but whether they took advantage – in an evolutionary sense – of changes in the distribution of these plants at the end of the Cretaceous is unclear. However, it is these Late Cretaceous patterns that need further investigation – from the point of view of both diversity fluctuations and also the ebb and flow of features that relate to feeding in herbivores and resistance to predation by plants.

Finally, species occurrences in the fossil record imply, via their phylogeny, unseen aspects of diversity that are reflected in their ghost lineages. Thus ghost lineages provide information on clade diversity beyond that available from raw species counts and help provide a clearer picture of minimal diversity levels, in this case among dinosaurian herbivores.

Acknowledgments

We thank H.-D. Sues for inviting us to contribute to this volume. We also thank L. M. Witmer, M. B. Meers, R. E. Heinrich, Z. Csiki, and M. A. Norell for discussions and comments on the ideas contained in this study.

At one stage or another, this research was supported by grants from the National Science Foundation (INT-8619987, EAR-8719878, EAR-9004458, SBR-9514784), the National Geographic Society, the National Research Council, and The Dinosaur Society.

References

Bakker, R. T. (1978). Dinosaur feeding behaviour and the origin of flowering plants. *Nature* 274:661–663.

Barrett, P. M. and Upchurch, P. (1994). Feeding mechanics of *Diplodocus*. *GAIA* 10:195–204.

Benton, M. J. (1984). Tooth form, growth, and function in Triassic rhynchosaurs (Reptilia: Diapsida). *Palaeontology* 27:737–776.

Calvo, J. O. (1994). Jaw mechanics in sauropod dinosaurs. *GAIA* 10:183–194.

Coe, M. J., Dilcher, D. L., Farlow, J. O., Jarzen, D. M., and Russell, D. A. (1987). Dinosaurs and land plants. In *The Origins of Angiosperms and Their Biological Consequences*, ed. E. M. Friis, W. G. Chaloner, and P. R. Crane, pp. 225–258. Cambridge and New York: Cambridge University Press.

Coombs, W. P., and Maryanska, T. (1990). Ankylosauria. In *The Dinosauria*, ed. D. B. Weishampel, P. Dodson, and H. Osmólska, pp. 456–483. Berkeley: University of California Press.

Crane, P. R. (1987). Vegetational consequences of angiosperm diversification. In *The Origins of Angiosperms and Their Biological Consequences*, ed. E. M. Friis, W. G. Chaloner, and P. R. Crane, pp. 107–144. Cambridge and New York: Cambridge University Press.

Crane, P. R. (1989). Patterns of evolution and extinction in vascular plants. In *Evolution and the Fossil Record*, ed. K. C. Allen and D. E. G. Briggs, pp. 153–187. Washington, DC: Smithsonian Institution Press.

Dodson, P. (1990a). Sauropod paleoecology. In *The Dinosauria*, ed. D. B. Weishampel, P. Dodson, and H. Osmólska, pp. 402–407. Berkeley: University of California Press.

Dodson, P. (1990b). Counting dinosaurs: how many kinds were there? *Proc. Natl. Acad. Sci. U.S.A.* 87:7608–7612.

Dodson, P., and Currie, P. J. (1990). Neoceratopsia. In *The Dinosauria*, ed. D. B. Weishampel, P. Dodson, and H. Osmólska, pp. 593–618. Berkeley: University of California Press.

Doyle, J. A., and Donoghue, M. J. (1986). Seed plant phylogeny and the origin of angiosperms: an experimental cladistic approach. *Bot. Rev.* 52:321–431.

Farlow, J. O. (1987). Speculations about the diet and digestive physiology of herbivorous dinosaurs. *Paleobiology* 13:60–72.

Fiorillo, A. R. (1991). Dental microwear on the teeth of *Camarasaurus* and *Diplodocus*: implications for sauropod paleoecology. In *Fifth Symposium on Mesozoic Terrestrial Ecosystems and Biota*, ed. Z. Kielan-Jaworowska, N. Heintz, and H. A. Nakrem, pp. 23–24. Oslo: University of Oslo.

Fleming, T. H., and Lips, K. R. (1991). Angiosperm endozoochory: were pterosaurs Cretaceous seed dispersers? *Am. Nat.* 138:1058–1065.

Forster, C. A. (1996). Species resolution in *Triceratops*: cladistic and morphometric approaches. *J. Vert. Paleont.* 16:259–270.

Forster, C. A., Sereno, P. C., Evans, T. W., and Rowe, T. (1993). A complete skull of *Chasmosaurus mariscalensis* (Dinosauria: Ceratopsidae) from the Aguja Formation (late Campanian) of west Texas. *J. Vert. Paleont.* 13:161–170.

Futuyma, D. J., and Slatkin, M. (eds.) (1983). *Coevolution*. Sunderland, MA: Sinauer.

Galton, P. M. (1985). Diet of prosauropod dinosaurs from the Late Triassic and Early Jurassic. *Lethaia* 6:67–89.

Galton, P. M. (1990). Basal Sauropodomorpha – prosauropods. In *The Dinosauria*, ed. D. B. Weishampel, P. Dodson, and H. Osmólska, pp. 320–344. Berkeley: University of California Press.

Hotton, N., III. (1986). Dicynodonts and their role as primary consumers. In *The Ecology and Biology of Mammal-Like Reptiles*, ed. N. Hotton, III, P. D. MacLean, J. J. Roth, and E. C. Roth, 71–82. Washington, DC: Smithsonian Institution Press.

Margulis, L., and Fester, R. (eds.) (1991). *Symbiosis as Source of Evolutionary Innovation*. Cambridge, MA: MIT Press.

Maryanska, T. (1990). Pachycephalosauria. In *The Dinosauria*, ed. D. B. Weishampel, P. Dodson, and H. Osmólska, pp. 564–577. Berkeley: University of California Press.

McIntosh, J. S. (1990). Sauropoda. In *The Dinosauria*, ed. D. B. Weishampel, P. Dodson, and H. Osmólska, pp. 345–401. Berkeley: University of California Press.

Niklas, K. J., Tiffney, B. H., and Knoll, A. H. (1980). Apparent changes in the diversity of fossil plants. *Evol. Biol.* 12:1–89.

Niklas, K. J., Tiffney, B. H., and Knoll, A. H. (1985). Patterns in vascular land plant diversification: an analysis at the species level. In *Phanerozoic Diversity Patterns*, ed. J. W. Valentine, pp. 97–128. Princeton: Princeton University Press.

Norell, M. A., and Novacek, M. J. (1993). Congruence between supraposititional and phylogenetic patterns: comparing cladistic patterns with the fossil record. *Cladistics* 8:319–337.

Norman, D. B. (1984). On the cranial morphology and evolution of ornithopod dinosaurs. In *The Structure, Development and Evcolution of Reptiles*, ed. M. J. W. Ferguson, pp. 521–547. *Symp. Zool. Soc. Lond.* 52.

Norman, D. B., and Weishampel, D. B. (1990). Iguanodontidae and related ornithopods. In *The Dinosauria*, ed. D. B. Weishampel, P. Dodson, and H. Osmólska, pp. 510–533. Berkeley: University of California Press.

Ostrom, J. H. (1964). A functional analysis of jaw mechanics in the dinosaur *Triceratops*. *Postilla* 88:1–35.

Sampson, S. D. (1995). Two new horned dinosaurs from the Upper Cretaceous Two Medicine Formation of Montana, with a phylogenetic analysis of the Centrosaurinae (Ornithischia: Ceratopsidae). *J. Vert. Paleont.* 15:743–760.

Sereno, P. C. (1986). Phylogeny of the bird-hipped dinosaurs (Order Ornithischia). *Natl. Geogr. Res.* 2:234–256.

Sereno, P. C. (1990). Psittacosauridae. In *The Dinosauria*, ed. D. B. Weishampel, P. Dodson, and H. Osmólska, pp. 579–592. Berkeley: University of California Press.

Sereno, P. C., and Dong, Z. (1992). The skull of the basal stegosaur *Huayangosaurus taibaii* and a cladistic analysis of Stegosauria. *J. Vert. Paleont.* 12:318–343.

Sues, H.-D., and Galton, P. M. (1987). Anatomy and classification of the North American Pachycephalosauria (Dinosauria: Ornithischia). *Palaeontographica* A 198:1–40.

Sues, H.-D., and Norman, D. B. (1990). Hypsilophodontidae, *Tenontosaurus*, Dryosauridae. In *The Dinosauria*, ed. D. B. Weishampel, P. Dodson, and H. Osmólska, pp. 498–509. Berkeley: University of California Press.

Swofford, D. L. (1985). *PAUP: Phylogenetic Analysis Using Parsimony*. Champaign, IL: Illinois Natural History Survey.

Thompson, J. N. (1994). *The Coevolutionary Process*. Chicago: University of Chicago Press.

Tiffney, B. H. (1986). Evolution of seed dispersal syndromes according to the fossil record. In *Seed Dispersal*, ed. D. R. Murray, pp. 273–305. North Ryde, NSW: Academic Press of Australia.

Upchurch, G. R., and Wolfe, J. A. (1987). Mid-Cretaceous – early Tertiary vegetation and climate: evidence from fossil leaves and wood. In *The Origins of Angiosperms and Their Biological Consequences*, ed. E. M. Friis, W. G. Chaloner, and P. R. Crane, pp. 75–105. Cambridge and New York: Cambridge University Press.

Upchurch, P. (1995). The evolutionary history of sauropod dinosaurs. *Phil. Trans. R. Soc. Lond. B* 349: 365–390.

Upchurch, P. (1998). The phylogenetic relationships of sauropod dinosaurs. *Zool. J. Linn. Soc.* 124:43–103.

Weishampel, D. B. (1984). Evolution of jaw mechanisms in ornithopod dinosaurs. *Adv. Anat. Embryol. Cell Biol.* 87: 1–110.

Weishampel, D. B. (1990a). Dinosaurian distributions. In *The Dinosauria*, ed. D. B. Weishampel, P. Dodson, and H. Osmólska, pp. 63–140. Berkeley: University of California Press.

Weishampel, D. B. (1990b). Ornithopoda. In *The Dinosauria*, ed. D. B. Weishampel, P. Dodson, and H. Osmólska, pp. 484–485. Berkeley: University of California Press.

Weishampel, D. B. (1996). Fossils, phylogeny, and discovery: a cladistic study of the history of tree topologies and ghost lineage durations. *J. Vert. Paleont.* 16:191–197.

Weishampel, D. B., and Horner, J. R. (1990). Hadrosauridae. In *The Dinosauria*, ed. D. B. Weishampel, P. Dodson, and H. Osmólska, pp. 534–561. Berkeley: University of California Press.

Weishampel, D. B., and Norman, D. B. (1989). The evolution of occlusion and jaw mechanics in late Paleozoic and Mesozoic herbivores. In *Paleobiology of the Dinosaurs*, ed. J. O. Farlow, Jr., pp. 87–100. *Geol. Soc. Am. Special Paper* 238.

Weishampel, D. B., and Witmer, L. M. (1990). Heterodontosauridae. In *The Dinosauria*, ed. D. B. Weishampel, P. Dodson, and H. Osmólska, pp. 486–497. Berkeley: University of California Press.

Weishampel, D. B., Norman, D. B., and Grigorescu, D. (1993). *Telmatosaurus transsylvanicus* from the Late Cretaceous of Romania: the most basal hadrosaurid. *Palaeontology* 36:361–385.

Wing, S. L., and Tiffney, B. H. (1987). The reciprocal interaction of angiosperm evolution and tetrapod herbivory. *Rev. Palaeobot. Palynol.* 50: 179–210.

Wing, S. L., Hickey, L. J., and Swisher, C. C. (1993). Implications of an exceptional fossil flora for Late Cretaceous vegetation. *Nature* 363: 342–344.

JOHN M. RENSBERGER

6

Dental constraints in the early evolution of mammalian herbivory

Introduction

The processing of plant food in mammals requires specialized mechanisms in order to sustain high rates of nutrient assimilation, and the evolution of herbivory in mammals can potentially be traced by examining changes in these structures. Conditions suitable for the preservation of soft tissues are exceedingly rare (Schaal and Ziegler 1993), and thus we cannot hope to trace the evolution of herbivory using structures of the stomach and intestinal tract. However, oral comminution of plant material is a nearly universal behavior among mammalian herbivores, owing to the problem of breaking down the plant cell walls composed of cellulose (Janis and Fortelius 1988). Fortunately, dental structures specialized for this purpose are well represented in the fossil record. One major group of Mesozoic mammals, the multituberculates, had already acquired masticatory mechanisms allowing herbivory (Hahn 1971; Krause 1982). However, the main mammalian radiation that led to modern herbivores probably occurred immediately following the extinction of the non-avian dinosaurs at the end of the Cretaceous. That event must have opened many niches that previously had been inaccessible to mammals, judging from the high taxonomic and morphological rates of evolution across the Cretaceous–Paleocene boundary (Sloan 1987).

This chapter examines the functional changes in the shapes of teeth in the two dominant groups of early Paleocene ungulates in North America, compares these with common Eocene ungulates, and considers the influence that tissue strength had on the directions of these changes. The results indicate that the changes were non-linear, apparently because of mechanical constraints that had to be overcome before the more efficient mastication characteristic of modern ungulates emerged.

Institutional abbreviations in this chapter are: MORV, Museum of the Rockies (fossil vertebrate collection), Bozeman, MT; UWBM, Burke Museum of Natural History and Culture, Seattle, WA. Other abbreviations: EDJ, enamel–dentine junction; SEM, scanning electron micrograph.

Tribosphenic mechanisms

The cheek teeth of the basal placental and marsupial insectivores that gave rise to the Cenozoic carnivores and herbivores have a tribosphenic cusp arrangement. Tribosphenic teeth have functional components that are found in later Cenozoic herbivores but the implementation and efficiencies of these functions are quite different.

Shearing mechanism The tribosphenic dentition consists of a complex arrangement of relatively sharp cusps that function to fracture food in several ways – through shearing, compressing or puncturing the food particles. These teeth have cusps with steeply inclined sides that slide past one another in a shearing movement during chewing. The teeth of the arctocyonid *Protungulatum*, a very primitive ungulate that occurs near the Cretaceous–Tertiary boundary (Sloan and Van Valen 1965), still retain much of the tribosphenic geometry, with cusps of moderately high aspect ratio and rather steeply inclined shearing surfaces (Figure 6.1). A fundamental part of this mechanism is a series of crests that extend along the sides of adjacent cusps (Figure 6.2). Like the cusps, the opposing occlusal edges of the crests glide past one another in close proximity during chewing movements (Figure 6.1).

These parts of the tribosphenic tooth, the cusp surfaces and crests, and the arrangement of the jaw muscles that move them can be called a shearing mechanism, but its effect on the food material is complex, producing tensile and compressive as well as shearing stresses. During the initial closing movement of the opposing tooth surfaces, the food is compressed. However, as the crest edges pass one another, the opposing edges push the food trapped between them in opposite directions, creating tensile and shearing stresses within the material, although these stresses are difficult to measure in the food.

If the food is brittle and not very fibrous, it may be fractured by the external compressive load before the edges pass one another in shear. When an object is compressed, tensile stresses are generated in directions at right angles to the direction of the compressive stresses (Rensberger

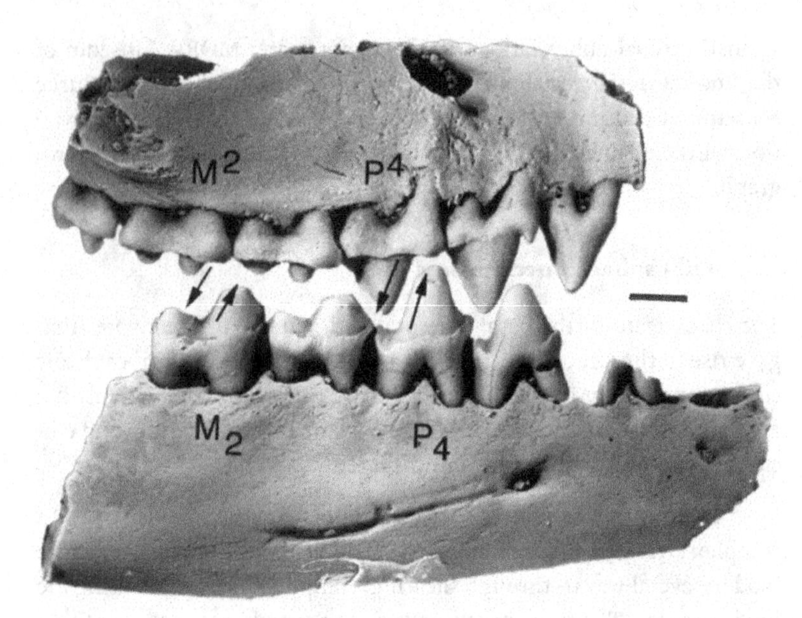

Figure 6.1. Buccal view of occluding P^2–M^3 and P_3–M_2 in the maxilla and dentary of arctocyonid *Protungulatum donnae* (MORV 5000 and 5001) from stream channel at top of the Upper Cretaceous Hell Creek Formation. Arrows indicate directions of shearing movement between opposing premolars and between crests of anterior side of upper molar and posterior side of lower molar trigonid. Scale bar (lower right of P^2) equals 2 mm.

1995: figs. 9.8, 9.16), and these tensile stresses may be high enough to produce fracture. For example, if the object is a seed with a hard outer shell and is subjected to vertical compression, vertical lines of fracture caused by horizontal tensile stresses will tend to form as the seed's diameter expands.

Puncture-crushing mechanism Observations of chewing activity in various living mammals (opossum, tree shrew, bat, primates, cat, and pig) indicate that a distinct type of mechanism, puncture-crushing, is used during early stages of processing a food object. The jaws are moved together in a rapid series of bites in which the opposing cusps do not reach full occlusion (Atkinson and Shepherd 1961; Ahlgren 1966; Crompton and Hiiemae 1970; Kallen and Gans 1972; Herring and Scapino 1973; Hiiemae and Kay 1973; Luschei and Goodwin 1974; Hiiemae and Thexton 1975). Puncture-crushing movements utilize the height and sharpness of the cusps, which open holes in the food material or fracture it if it is brittle.

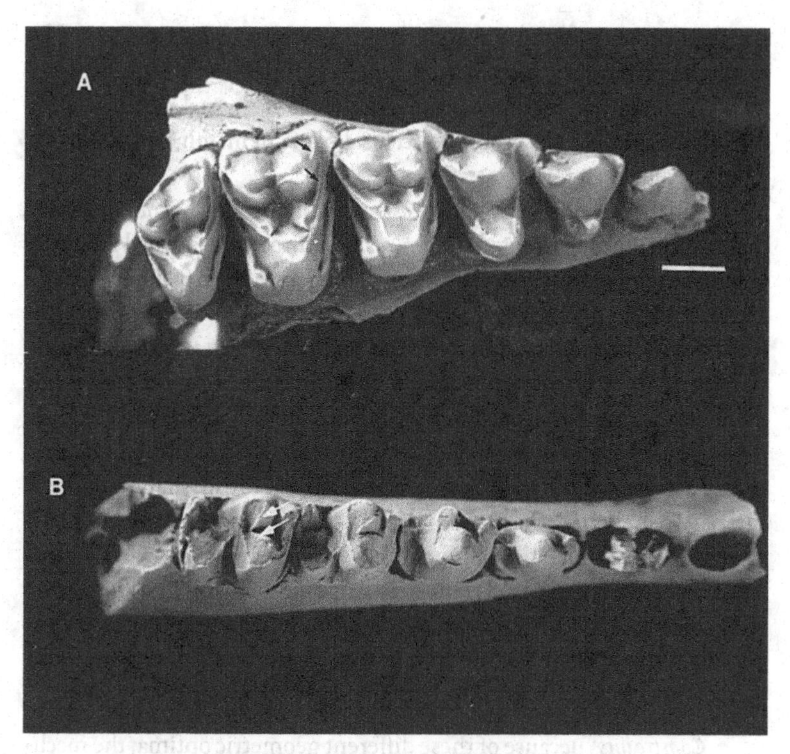

Figure 6.2. Occlusal views of (A) right P^2–M^3 (MORV 5000) and (B) right P_3–M_2 (MORV 5001) in *Protungulatum donnae*. Arrows point to crests on (A) anterior side of paracone that shear against crests on (B) posterior side of trigonid. Scale bar (occlusal view, lower right) equals 2 mm.

Compressive mechanism In addition to the compression generated by the previously described shearing mechanism, tribosphenic teeth have a structure specialized solely for compression, the talonid basin of the lower molar that opposes the tip of the protocone of the upper molar. The opposition of protocone tip against concave surface of the talonid basin efficiently flattens food caught between these structures, bursting membranes and cell walls.

Efficiency

A measure of efficiency of mechanisms for oral food processing is the amount of increased food surface area produced per chewing stroke. The increase in surface area for materials that are fractured by shearing activity is equivalent to twice the sum of the lengths of the cut edges multiplied by the depth of the occlusal stroke or thickness of the material (Rensberger 1973:520). Edge lengths are measured in the plane normal to the chewing direction – in other words, in the occlusal plane in mammals with orthal (vertical) chewing movement.

For small food objects, a compressive mechanism can be more efficient than a shearing one. In compression, a small three-dimensional food particle is flattened into a geometry approaching a two-dimensional surface. For example, a spherical object compressed to one-fifth its original diameter attains a surface of about twice the original surface area. By comparison, the same sphere cut in half by a shearing edge produces only about 1.5 times the original surface area. Frugivores, which utilize food consisting largely of non-fibrous materials with cells that are easily burst and spread across surfaces, thus tend to have relatively broad, shallow occlusal surfaces to maximize the amount of food that is compressed during each stroke. Primates (including humans) have relatively large occlusal basins and hence have emphasized the compressive mechanisms in tribosphenic teeth. Not unexpectedly, the rate of comminution of easily fractured food materials has been found to correlate in humans more strongly with occlusal area than with dental length measures (Duke and Lucas 1985).

Constraints Because of these different geometric optima, the mechanisms for shearing and compression compete for space in a given dentition. Furthermore, because efficiency in shearing mechanisms depends on total edge length, efficiency in mammals with vertical chewing movements is limited by the width (or length) of the tooth and the number of crests that can be arranged across that space. Each crest must lie adjacent to a space for receiving the occluding crest on the opposing tooth, further limiting the number of crests that can occupy a given area. Crest density is also limited by the strength of the dental materials: thinner crests fracture under lower stresses than thicker crests composed of the same material. Consequently, larger mammals with orthal chewing and greater bite force require absolutely thicker crests and intervening spaces and cannot simply pack an increased number of thin crests into the space available in a given jaw size.

Changes in occlusal geometry

The evolution of herbivory in Cenozoic mammals must have faced these constraints, and new occlusal geometries that appeared may be closely

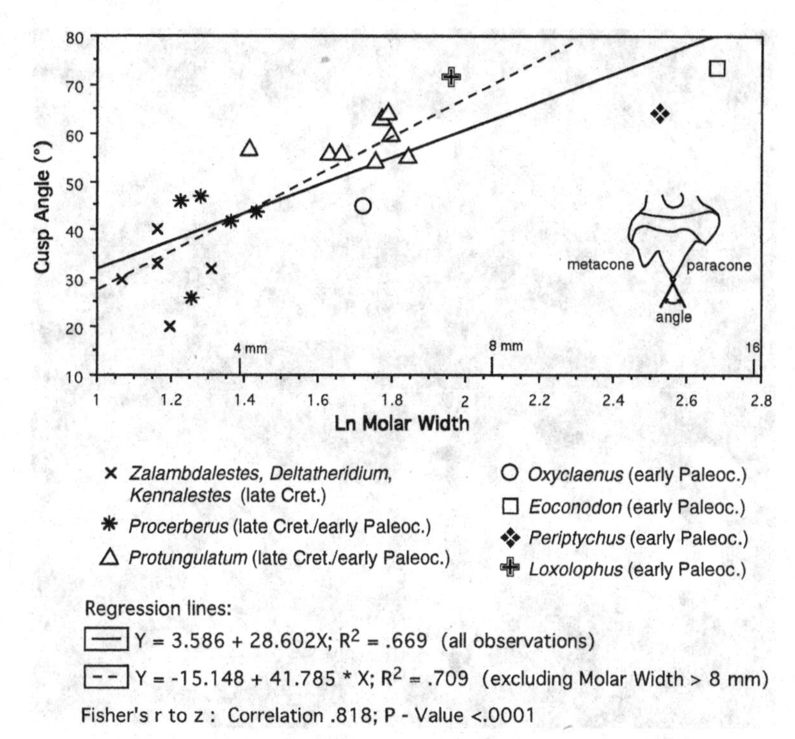

Figure 6.3. Relationship of taper angle of paracone on M^2 with molar width (as indicator of relative body size) in Late Cretaceous and early Paleocene insectivores and ungulates. In some specimens of *Procerberus* and *Protungulatum* the tooth may be an M^1.

tied to them. One conspicuous change in occlusal geometry found in both major groups of early Paleocene ungulates, arctocyonids and periptychids, was a reduction in cusp sharpness.

Relationship of cusp bluntness to body size How cusp sharpness varies among taxa can be examined by measuring the angle formed by the slopes of the surfaces on either side of a cusp tip. When cusp angle and tooth size are compared across a range of Late Cretaceous and early Paleocene mammals (Figure 6.3), it is apparent that cusp shapes become blunter as size increases.

In the early arctocyonid *Protungulatum* (Figures 6.1–6.2), the molar cusps are already of somewhat lower aspect ratio than in Cretaceous insectivorous taxa. In a specimen of the larger and stratigraphically slightly younger arctocyonid *Eoconodon* (Figure 6.4), the protoconids are lower in height relative to their basal widths than those in *Protungulatum*.

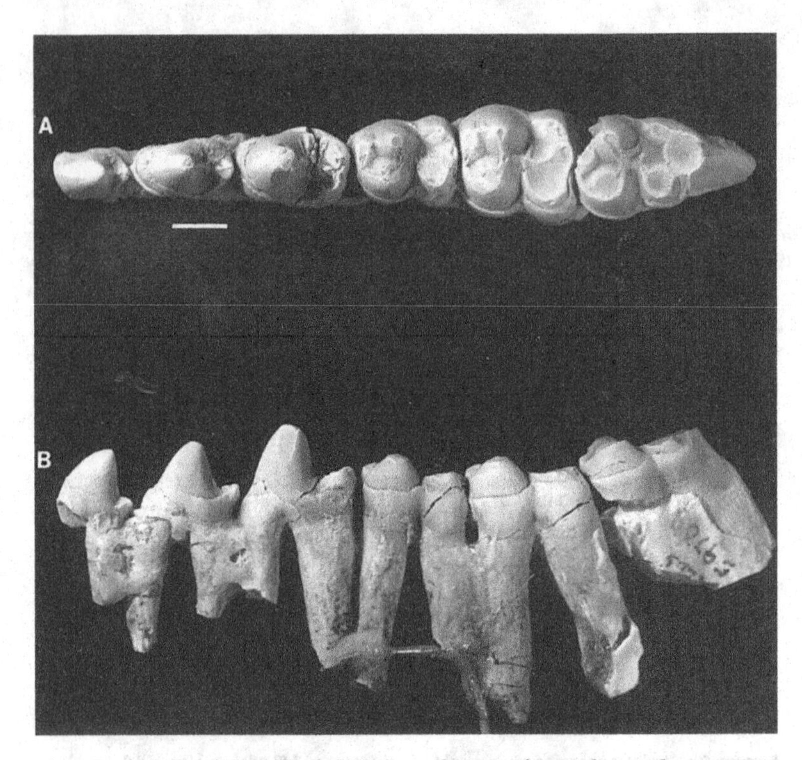

Figure 6.4. Occlusal (A) and buccal (B) views of P2–M3 of *Eoconodon coryphaeus*, UWBM 59709, from the base of the lower Paleocene Nacimiento Formation (Puercan). Scale bar (occlusal view, lower left) equals 5 mm.

Another change was the reduction of occlusal crests. Crests still connect the molar cusps in *Protungulatum* and functioned as edges that sheared against crests of the opposing teeth (Figure 6.2), but these are greatly reduced in *Eoconodon* (Figure 6.4A) to the extent that the occlusal surface is more strongly dominated by cusps of circular cross-section. The replacement of crests by circular cusps is even more pronounced in the periptychid condylarths of early Paleocene age. Typical adult dentitions of *Ectoconus* and *Periptychus* have the cusp tips worn away to an extreme extent, forming broad, flat surfaces (Figure 6.5) dominated by circular enamel borders. The cusp shapes in *Eoconodon* and the periptychids contrast with those of most later Cenozoic ungulates, which have teeth dominated by linear edges associated with more horizontal chewing directions (Rensberger 1973, 1986, 1988). An examination of the direction of the striae on the wear surfaces of periptychids indicates a strictly orthal

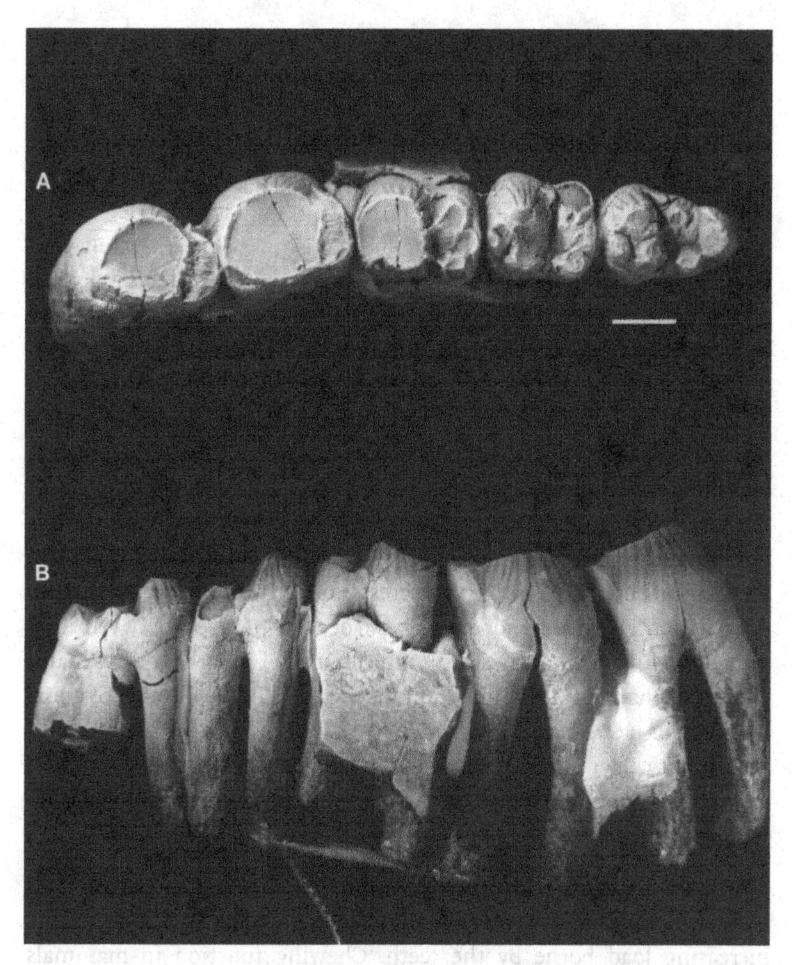

Figure 6.5. Occlusal (A) and buccal (B) views of right P3–M3 of *Periptychus coarctatus*, UWBM 59686, from base of Nacimiento Fm., Puercan, early Paleocene. Scale bar (occlusal view, lower right) equals 5 mm.

(vertical) chewing direction, lacking the horizontal movements that characterize most later ungulates (Rensberger 1986).

This association between cusp bluntness, body size, and the loss of crests may be explainable as responses to the relative efficiencies of compressive and shearing mechanisms, the physiological demands of increasing body size, and the strength of the dental materials. As early ungulates became larger, the primitive, primarily orthal chewing motion remained unmodified. Because of the different geometric optima of the mechanisms for compression and shearing, the compressive mechanism was more efficient (as noted above) under vertical chewing movements and, if tender plant parts were available in sufficient quantity, could meet the increasing nutritional demands of larger bodies if the diameters of the compressive areas of the teeth simply increased in proportion to the increase in linear dimensions of body size.

On the other hand, the shearing mechanism was more severely constrained as body size increased because the additional crest length needed to keep pace would have required either radical changes in the crest geometry and chewing direction or disproportionate increases in tooth size. For example, simply increasing the height of a cusp, which would increase the length of a crest running from the tip to the base of the cusp, does not increase the effective length of the crest, which is measured in a direction perpendicular to the shearing direction. Crest lengths in tribosphenic teeth have as their upper limits the more or less horizontal distances between major cusps, which are already situated on the margins of the teeth. For the crests to function in shearing, there must be space for the opposing crests to occlude precisely between adjacent crests, and generating crests in the interior of a tooth requires major geometric changes. Although dramatic increases in the number of crests do occur in later Cenozoic herbivores, they are always associated with horizontal chewing movements in which 'grinding' occurs parallel to the occlusal plane and which thus avoid the need for intercrest occlusion with opposing crests.

Load In addition to the constraints on shearing mechanisms as body size increased, another factor limiting the avenues of change was the increasing load borne by the teeth. Chewing function in mammals depends on convergence in small occlusal areas of the forces generated by the jaw musculature, and larger body size results in larger muscles and larger forces. Without reduction in the stress-concentrating mechanisms of sharp cusps and crests, this would result in much larger stresses within the dental materials.

An additional advantage of enlarged compressive regions in the teeth is that the brittle materials forming the teeth are much stronger in response to compressive than to tensile stresses. The broader and flatter compressive surfaces generate lower tensile stresses than narrower, sharper structures (see below). This may be the most important factor influencing the early evolutionary modifications of cusp shape.

One can see the effect of stresses on cusps of lower height-to-breadth

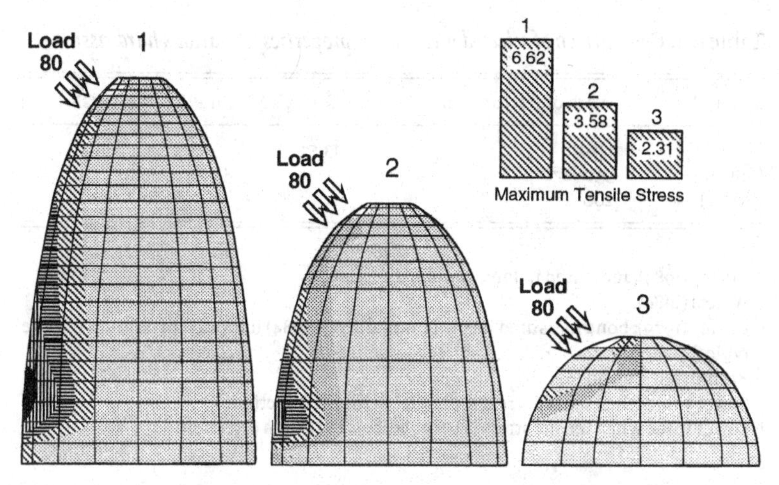

Figure 6.6. Relationship of cusp shape to stress magnitude in enamel. Finite element models of tooth cusps of different heights but same base size, subjected to the same load. Outer layer of elements of model has higher Young's Modulus than inner elements to simulate enamel layer surrounding dentine.

ratios by modeling the stress regimes in cusps of different aspect ratios while keeping the loading forces and diameters of the cusps constant. Figure 6.6 illustrates three finite-element models of cusps of different heights but with the same basal cross-section and subjected to the same load. As the cusp acquires a lower aspect ratio, the maximum tensile stresses decrease. Reduction of aspect ratio (increasing cusp bluntness) and reduction in crests (structures that increase aspect ratios in small areas) is a simple way to minimize the tensile stresses within the teeth as body and muscle size increase.

Cusp bluntness and dietary behavior Although blunt cusps increase the amount of surface normal to the vertical chewing direction and thus have the effect of enhancing the efficiency of the compression mechanism, selection in this direction at the same time increasingly restricts the range of food materials that can be efficiently processed because fibrous plant tissues require shearing mechanisms for comminution. Owing to the limited availability of easily processed, soft plant parts that are low in cellulose, as body size continued to increase and demand larger volumes of nutrients, the selective advantage of any innovation that would allow the acquisition of shearing mechanisms and higher occlusal stresses must have been considerable.

Table 6.1. *Comparison of selected mechanical properties for various hard tissues*

Material	Work of fracture (J m^{-2})	Young's Modulus (stiffness) (GN m^{-2})
Bone	1710[c]	13.5[a]
Dentin	550–270[d e]	11–16[b]
Enamel	200–13[d f]	20–90[b]

Notes:
[a] Currey (1984), measured in longitudinal direction.
[b] Waters (1980).
[c] Currey (1979); bone measured is cow femur. Currey (1984) indicated maximum is more typically 2800.
[d] Rasmussen *et al.* (1976).
[e] Lower value when tensile stress is parallel to tubule direction.
[f] Lower value when tensile stress is perpendicular to prism direction.

Enamel reinforcement

Acquisition of stronger enamel appears to have been one of those innovations that allowed elevation of occlusal stresses and evolution of new shearing mechanisms. Enamel is probably the densest and most brittle material of the skeleton and consequently most easily fractured (see Table 6.1). The only other material in the mammalian skeleton that approaches the weakness of enamel to tensile stresses is the dense bone of the ear region, which is seldom subjected to high loads.

Fracture propagation To understand how enamel could be restructured to become more resistant to stresses, one must consider how fractures propagate. Fracture in stiff, brittle materials is promoted by the increasing magnitude of stress concentrated at the tip of a propagating crack. As a crack lengthens, an increasing proportion of the tensile stress trajectories in the material bends around the crack tip, increasing the stress concentration at the tip (e.g., Rensberger 1995: fig. 9.2).

It is clear that a microstructural condition that reduces the concentration of stress at the crack tip would help prevent the fracture from increasing in length. Prism decussation (zones of prisms with a common orientation alternating with zones of prisms with a different orientation) is a common feature of mammalian enamel and accomplishes this goal in several ways.

Several attributes of the stress field surrounding a propagating crack can be exploited by the microstructure to inhibit crack expansion. The stress concentration at the crack tip varies directly with the sharpness of the tip. In addition to the stresses parallel to the nominal forces acting on a tooth or other object, other stresses aligned in directions normal to the nominal stress direction exist ahead of the tip of a propagating crack. Because prism boundaries are microstructurally the weakest regions in prismatic enamel, if prisms running oblique to the main crack plane exist in front of a propagating crack, these 'ahead' stresses will tend to open small cracks between the obliquely running prisms in front of the main crack tip. As the main crack reaches these oblique cracks, the latter have the effect of enlarging the tip of the main crack and thereby reducing the stress concentration (Rensberger 1997: fig. 4) or simply causing the crack to turn and follow the new interface direction. That direction is less favorable to crack propagation because it is less normal to the stresses acting to pull the crack faces apart (Pfretzschner 1988).

There are other factors associated with prism decussation that inhibit crack propagation (Rensberger 1995), including the complexity of the new surface generated by the crack. Decussating prisms cause fractures passing through the enamel to deviate from the main plane of the fracture and form larger, more rugose surfaces. Regardless of the material, new surface generated by cracks subtracts energy and tends to inhibit further propagation (Griffith 1921).

Enamel structure in early placental herbivores Although prism decussation is common among Cenozoic mammals, it was a rare attribute in the earlier Paleocene forms (Koenigswald, Rensberger and Pfretzschner 1987). Periptychid condylarths, which were abundant during the early Paleocene, lacked it altogether. A sample of enamel from the early Paleocene *Hemithlaeus* shows prisms extending in parallel from the enamel–dentine junction (EDJ) to the outer surface of the tooth without bending at all (Figure 6.7).

Prisms in a much larger early Paleocene periptychid, *Periptychus*, are also uniformly parallel throughout their course, and extend from the EDJ in the same outward–upward direction (Figure 6.8). There is, however, a narrow interval in the inner part of the enamel, at about one-fifth the distance to the outer surface, where the prisms bend cervically as well as out of the radial plane (i.e., cervically and circumferentially). In Figure 6.8, the greater electron emission (lighter gray color) on one side of the bend is due to the out-of-plane slope of the fracture surface. Although this is not prism decussation, this sigmoidal bend is probably a crack-resisting mechanism because a crack plane cannot remain perfectly flat while passing through the enamel along the easiest path, the straight route fol-

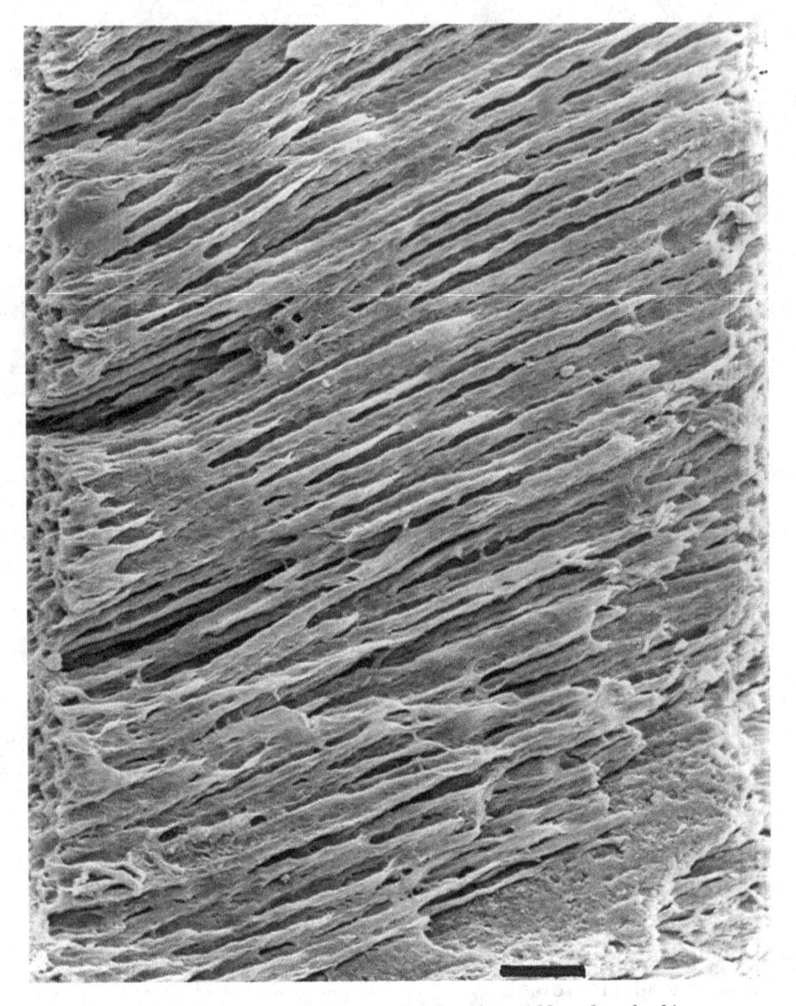

Figure 6.7. Enamel structure of a lower premolar of *Hemithlaeus kowalevskianus*, UWBM 59469, Puercan (early Paleocene). SEM (scanning electron micrograph) of lightly etched fractured surface. EDJ (enamel–dentine junction) at left, outer surface at right, occlusal direction toward top. Scale bar (near lower right) equals 20 μm.

lowing prism boundaries. In addition to forcing the crack to follow a less favorable direction, the out-of-plane bend slightly increases the surface area of the crack plane and thus helps resist the growth of the crack. The vertical continuity of this bend and the extreme uniformity of the prism direction elsewhere in the enamel are not known in any later mammal of this or larger size. This appears to be a very simple acquisition that slightly improved the strength of the enamel.

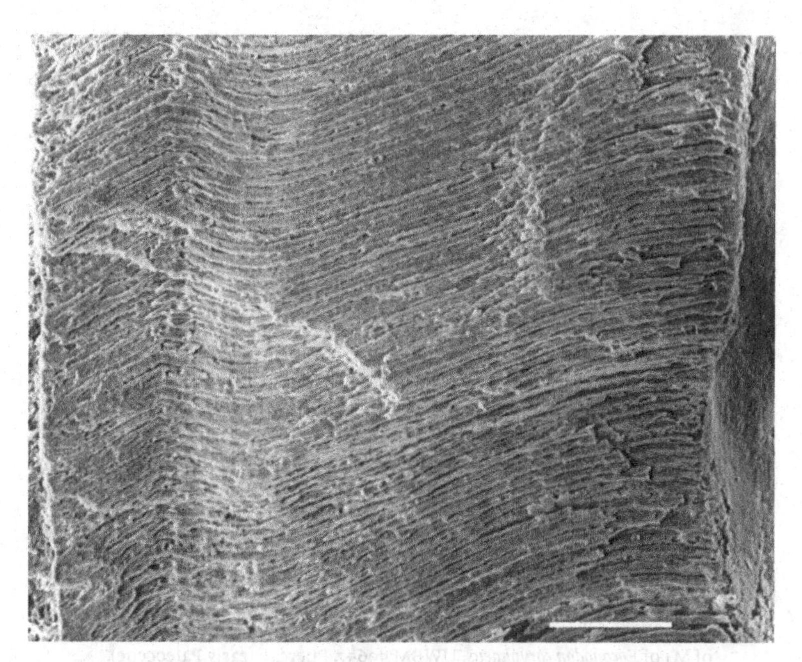

Figure 6.8. SEM of the structure of the buccal enamel of a P4 of *Periptychus coarctatus*, UWBM 59671, Puercan (early Paleocene). Occlusal direction toward top, EDJ at left. Scale bar (lower right) equals 50 μm.

Small divergences in prism direction also occur in the early Paleocene arctocyonid *Eoconodon* (Figure 6.9). Most prisms tend to arise at the EDJ and follow almost the same occlusolateral direction as in the periptychids. However, some bundles of prisms occupying regions ranging up to 100 μm or more in diameter bend in a slightly cervical or slightly circumferential direction. The diverging prisms sometimes maintain the new direction until they reach the outer surface. This decussation is very rudimentary. The diverging prisms usually form small angles with neighboring prisms of about 30° to 40° and the decussation is scattered and irregular. The small angle and the irregularity contrasts with the larger angles of decussation and regular spacing of decussating zones found in extant medium-sized or larger mammals (Kawai 1955; Koenigswald 1997).

Specimens of *Loxolophus*, another early Paleocene arctocyonid, show a slightly greater degree of regularity in prism decussation. In UWBM 59725, the nominal prism direction forms an angle of about 50° with the EDJ, but there are bundles that run more directly outward (angled about 70° or so with the EDJ). The interspaces between these divergent bundles

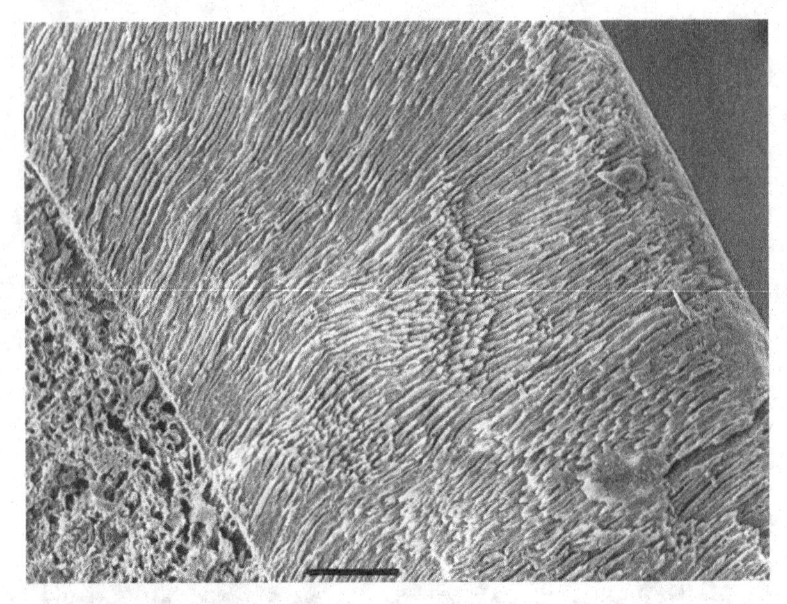

Figure 6.9. SEM of the structure of the enamel on the lingual side of the metaconid of M1 of *Eoconodon coryphaeus*, UWBM 59657, Puercan (early Paleocene). Occlusal toward upper left, outer surface at right, and dentine and EDJ at lower left. Scale bar (bottom, left of center) equals 50 μm.

are sometimes subequally spaced along the occlusal–cervical axis of the tooth (Figure 6.10). This pattern, though not consistently developed, is suggestive of those in more advanced mammals with clearly decussating prisms. In some regions (Figure 6.10), the widths of the decussating bundles of prisms are narrow, but the widths are 130 μm or more in others (Figure 6.11). In some regions no repeating pattern is evident. The angle of divergence in prism direction is typically 30° to 40°.

Enabling effect of prism decussation

At certain body sizes, the available volume of easily processed plant parts, such as fruits and seeds, would no longer be sufficient to sustain growth and maintenance functions, even though they are rich in digestible nutrients. At that point, occlusal mechanisms allowing access to the tougher but more abundant and evenly distributed plant parts would be required. Blunt teeth with only simple compressive mechanisms and lacking crests would be unable to process efficiently the required volumes of the more fibrous plants. In order to attain

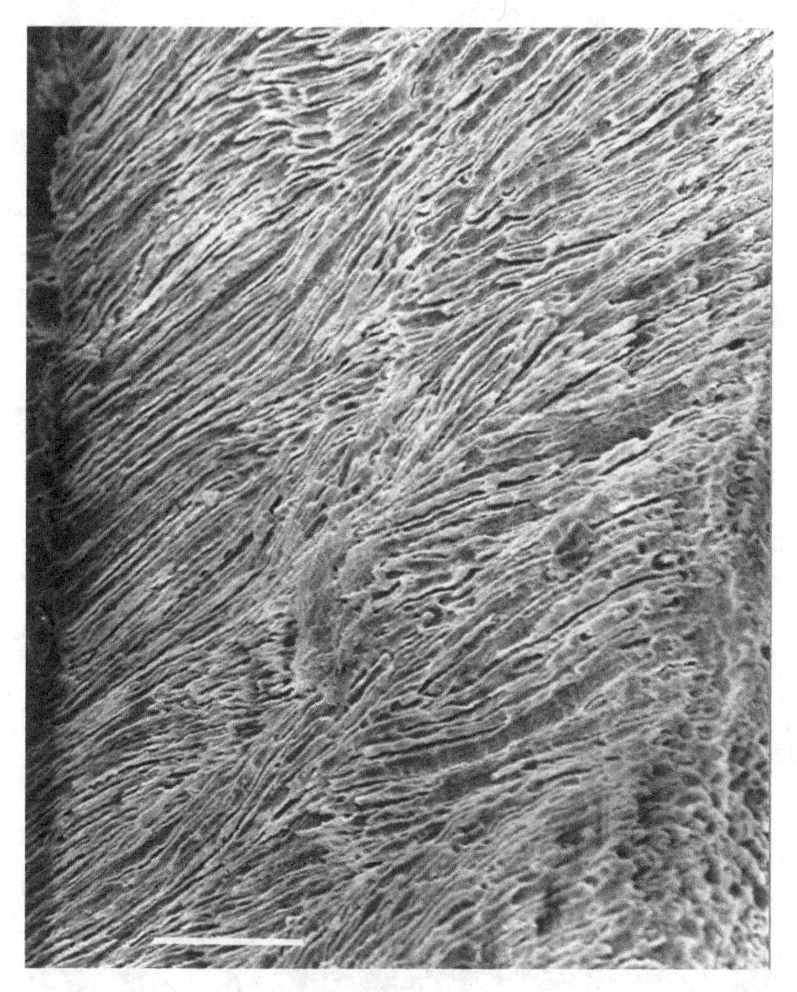

Figure 6.10. Enamel structure of upper premolar of *Loxolophus hyattianus*, UWBM 59725, Puercan (early Paleocene). SEM of section near tip of paracone, EDJ at left, tip of cusp toward top, outer surface at right. Scale bar (lower left) equals 50 μm.

sharper occlusal structures, a stronger enamel reinforcement than those present in the early Paleocene mammals discussed above would become necessary.

When data from Eocene ungulates are added to the graph of cusp angle (Figure 6.12), the smaller of these later taxa, such as *Homogalax*, *Hyracotherium*, and *Ectocion*, fall close to the regression line for Cretaceous and Paleocene taxa (Figure 6.3). However, the larger of the Eocene and

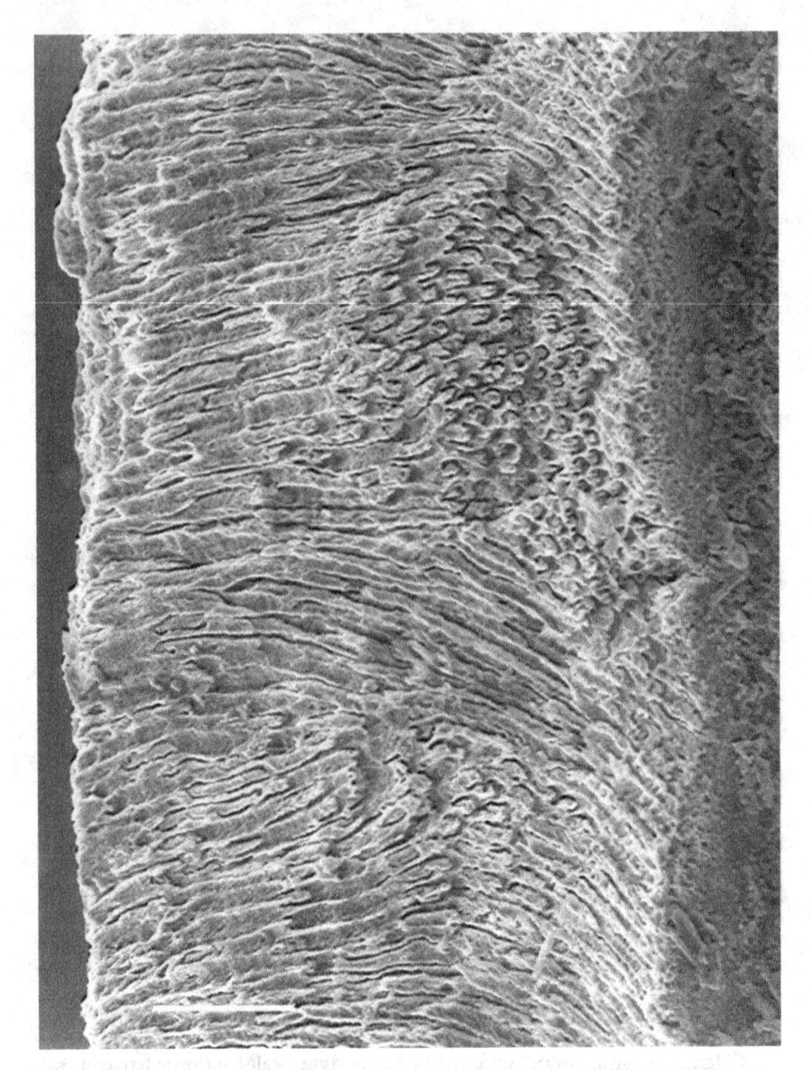

Figure 6.11. Enamel structure of upper premolar of *Loxolophus hyattianus*, UWBM 59725, Puercan (early Paleocene). SEM of section 2 mm below tip of paracone on opposite side of cusp from Figure 6.10. EDJ at right, occlusal toward top, outer surface at right. Scale bar (lower left) equals 50 μm.

Oligocene taxa returned to small cusp angles and appear to be free of the body-size constraint. The narrow angles in the occlusal structures of the large forms like *Coryphodon* (Figure 6.13), together with the extensive jaw musculature required to move such massive jaws, indicates that high peak stresses were occasionally concentrated in the occlusal structures. Nevertheless the steep sides of the lophs and cusps were critical in main-

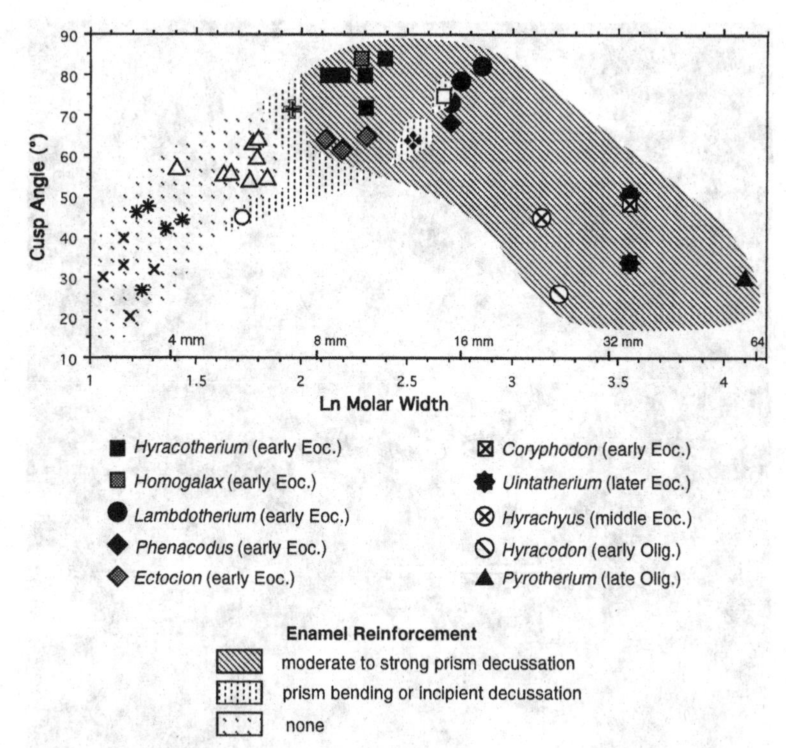

Figure 6.12. Relationship of cusp angle to width of upper molar in Eocene ungulates contrasted with Cretaceous/early Paleocene taxa. See also definitions of symbols in Figure 6.3. The cusp angle was measured on the paracone of M^2 except that where the paracone is diminished or merged into a loph, the following was substituted: *Hyrachyus* and *Hyracodon* (taper angle of ectoloph near metacone); *Coryphodon*, *Pyrotherium* (taper angle of metacone). Two angles were measured on *Uintatherium*: the larger angle was measured on the paracone, the smaller angle on the metacone.

taining sharp edges on these shearing structures (Figure 6.13) as the occlusal surfaces wore down.

A key factor that seems to have facilitated the reacquisition of shearing mechanisms was toughening of the brittle enamel by prism decussation. Prism decussation is much better developed in the Eocene ungulates than in the early Paleocene taxa. The only two ungulates illustrated in Figure 6.12 with molar widths greater than 7 mm that lack moderate to strong prism decussation are *Periptychus* and *Eoconodon*, but these taxa lived during the early Paleocene, at a time when no mammals had acquired anything but incipient enamel reinforcement.

It is noteworthy that none of the ungulates shown in Figure 6.12 had acquired the strong translatory chewing movements that characterize

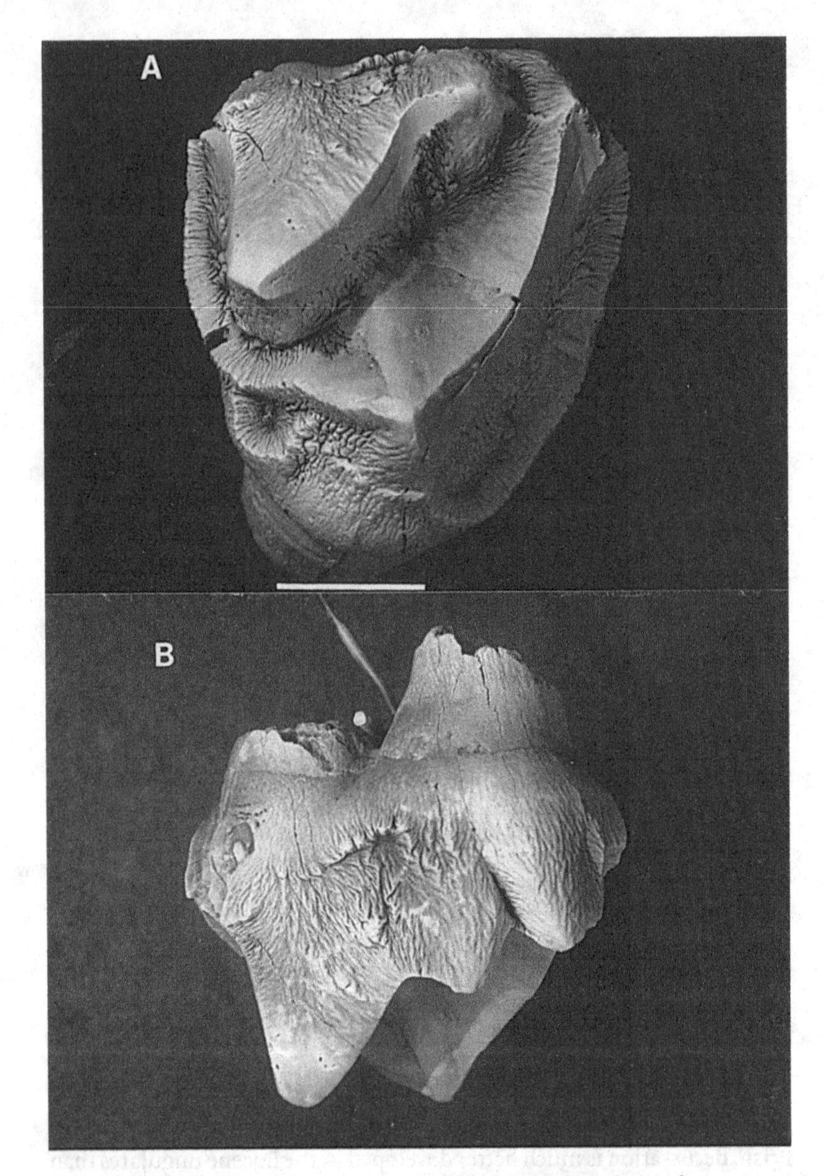

Figure 6.13. Occlusal (a) and buccal (b) views of linear shearing edges and narrow cusp angles on an upper molar of *Coryphodon* sp., UWBM 59329, Wasatchian (early Eocene). Scale bar (below occlusal view) equals 10 mm.

most successful ungulate lineages of the later Cenozoic. *Coryphodon, Uintatherium,* and *Pyrotherium* retained the strongly orthal chewing movement of primitive mammals. However, the rather transverse lophs of the lower occlusal surfaces struck the lophs of the upper surfaces off-center, causing shearing along a plane making an angle of roughly 45° with the vertical direction. (Note the inclined angles of the shearing surfaces on the lophs of *Coryphodon*; Figure 6.13.) The shearing surfaces in *Pyrotherium* and *Uintatherium* are mechanically similar.

The protoloph and metaloph of the upper molars in the early rhinocerotoids *Hyrachyus* and *Hyracodon* are also rather transverse, yet these lophs do not interdigitate between the lophs of the upper dentition but move across them in nearly horizontal planes. The ectoloph forms a steep anteroposteriorly aligned shearing mechanism that functioned during the early phase of jaw closure (Rensberger 1992: fig. 8).

During the later Cenozoic, horses and various selenodont artiodactyls exploited the advantages of predominantly transverse chewing movements, which allowed them to employ multiple, closely spaced enamel edges and heavy emphasis on shearing for more efficient processing of fibrous plant fodder. Interestingly, enamel reinforcement in the advanced horses and artiodactyls is accomplished to a great extent by a different type of microstructure (Figure 6.14; Pfretzschner 1992, 1993; Kozawa and Suzuki 1995). This enamel functionally consists of thin, flat sheets of hydroxyapatite crystallites, which in alternating laminae extend in opposing directions. In the modern horse, the prisms, which form each alternate layer, have become so flattened and aligned so completely in the same plane, and the interprismatic matrix has become so similarly flattened as an alternating layer, that almost the only distinction between prismatic and interprismatic enamel is the roughly 75° difference in alignment of the crystallites in the alternating layers (Rensberger 1996). This type of enamel is the most plywood-like of all known hard tissues in vertebrates.

Summary and conclusions

Hadrosaurid dinosaurs and late Cenozoic ungulate mammals, including horses and bovids, had elaborate dental mills for processing vegetation at high rates. However, the evolution of dental structures in early ungulates was impeded by constraining factors related to their derivation from those of insectivorous mammals.

The masticatory system in insectivorous mammals utilizes high,

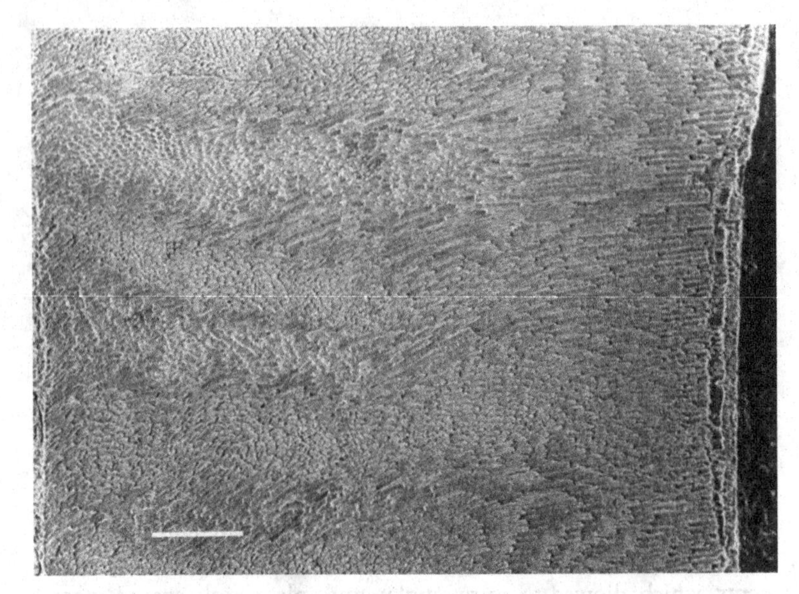

Figure 6.14. Enamel structure of M3 of *Phenacodus primaevus*, UWBM 39030, Wasatchian (early Eocene). SEM of polished and etched radial section, lingual side of metaconid. EDJ at left margin of micrograph, outer surface at right, occlusal toward top. Prism decussation zones (lighter and darker bands) run outward and slightly cervically from near the EDJ, but disappear in outer third of enamel thickness. Scale bar (lower left) equals 100 μm.

sharp cusps that interdigitate with opposing cusps during largely vertical chewing movements. These complex teeth contained three mechanisms for fracturing food. The insertion of cusps between and against opposing cusps provides a shearing mechanism, but, at the same time, the movement of the protocone of the upper molars into the talonid basin in the lowers adds a compressive mechanism, and the height and sharpness of the cusps adds a puncture-crushing mechanism.

In the early Paleocene ungulates, the shearing component was greatly reduced as the ratio of cusp height to cusp width was reduced and crests were lost. These changes seem to provide maximal functional efficiency, given the limitation of space, and made compression the dominant mechanism. The compressive mechanism, which depends on areas of surface in the plane normal to the chewing direction, would have been able to keep up with the increased nutrient requirements of larger body size by increasing its linear dimension (radius) in proportion to linear increases in body size, whereas the shearing mechanisms, which depend on lengths of crests, would have needed to increase at a higher rate.

A second factor influencing the shift toward compressive mechanisms appears to have been enamel strength. Modeling shows that the stresses in tooth enamel induced by chewing activity are lower in low, wide cusps than in high, sharp cusps. It is clear that prism decussation makes the enamel more resistant to fracture, and, by early Eocene times, moderate to strong prism decussation had appeared in most ungulates with upper molars of at least 7 mm in width. From this size upward in early Eocene ungulates, the earlier trend toward cusp bluntness is reversed, occlusal structures again have higher aspect ratios, and crests reappear.

The early increase in cusp angle and loss of crests with increased body size, followed by later reduction of cusp angle after prism decussation became established in larger mammals, implies that enamel brittleness was an important constraint on the evolution of herbivory during the early Cenozoic. This constraint, together with the different mechanics of shearing and compressive mechanisms, must have combined to limit the dietary diversity and/or the rates of nutrient intake in many early Cenozoic herbivores.

Acknowledgments

I am grateful to Dr Hans-Dieter Sues for inviting me to participate in the symposium on the evolution of herbivory and to Drs Donald L. Rasmussen and John R. Horner who, respectively, collected and lent the fine jaws of *Protungulatum donnae* for this study. Drs Mikael Fortelius, John Hunter, and Tracy Popowics kindly reviewed the manuscript and contributed comments and suggestions that resulted in its significant improvement. Beverly Witte prepared the Paleocene and Eocene fossil material from the UWBM collection, and she, Ben Witte, and Mesfin Asnake collected many of the specimens used in this project.

References

Ahlgren, J. (1966). Mechanisms of mastication. *Acta Odont. Scand.* 24:5–109.

Atkinson, H. F., and Shepherd, R. W. (1961). Temporomandibular joint disturbances and the associated masticatory patterns. *Aust. Dent. J.* 6:219–222.

Crompton, A. W., and Hiiemae, K. M. (1970). Molar occlusion and mandibular movements during occlusion in the American opossum, *Didelphis marsupialis* L. *Zool. J. Linn. Soc.* 49:21–47.

Currey, J. D. (1979). Mechanical properties of bone with greatly differing functions. *J. Biomechanics* 12:313–319.

Currey, J. D. (1984). *The Mechanical Adaptations of Bones*. Princeton: Princeton University Press.

Duke, D. A., and Lucas, P. W. (1985). Chewing efficiency in relation to occlusal and other variations in the human dentition. *British Dent. J.* 159:401–403.

Griffith, A. A. (1921). The phenomena of rupture and flow in solids. *Phil. Trans. R. Soc. Lond. A* 221:163.

Hahn, G. (1971). The dentition of the Paulchoffatiidae (Multituberculata, Upper Jurassic). *Mem. Serv. Geol. Portugal (N.S.)* 17:1–39.

Herring, S. W., and Scapino, R. P. (1973). Physiology of feeding in miniature pigs. *J. Morph.* 141:427–460.

Hiiemae, K. M., and Kay, R. F. (1973). Evolutionary trends in the dynamics of primate mastication. In *Craniofacial Biology of the Primates*, ed. M. R. Zingeser, pp. 28–64. *Symposium 4th Int. Congr. Primat.* Basel: Karger.

Hiiemae, K. M., and Thexton, A. J. (1975). Consistency and bite size as regulations of mastication in cats. *J. Dent. Res.* 54(A):194.

Janis, C. M., and Fortelius, M. (1988). On the means whereby mammals achieve increased functional durability of their dentitions, with special reference to limiting factors. *Biol. Rev.* 63:197–230.

Kallen, F. C., and Gans, C. (1972). Mastication in the little brown bat (*Myotis lucifugus*). *J. Morph.* 136:385–420.

Kawai, N. (1955). Comparative anatomy of the bands of Schreger. *Okimimas Folia Anat. Jap.* 27:115–131.

Koenigswald, W. von. (1997). Brief survey of enamel diversity at the schmelzmuster level in Cenozoic placental mammals. In *Tooth Enamel Microstructure*, ed. W. von Koenigswald and P. M. Sander, pp. 137–161. Rotterdam: Balkema.

Koenigswald, W. von, Rensberger, J. M., and Pfretzschner, H.-U. (1987.) Changes in the tooth enamel of early Paleocene mammals allowing increased diet diversity. *Nature* 328:150–152.

Kozawa, Y., and Suzuki, K. (1995). Appearance of new characteristic features of tooth structure in the evolution of molar teeth of Equoidea and Proboscidea. In *Aspects of Dental Biology: Palaeontology, Anthropology and Evolution*, ed. J. Moggi-Cecchi, pp. 27–31. International Institute for the Study of Man.

Krause, D. W. (1982). Jaw movement, dental function, and diet in the Paleocene mulituberculate *Ptilodus*. *Paleobiology* 8:265–281.

Luschei, E. S., and Goodwin, G. M. (1974). Patterns of mandibular movement and jaw activity during mastication in the monkey. *J. Neurophysiol.* 37:954–966.

Pfretzschner, H.-U. (1988). Structural reinforcement and crack propagation in enamel. In *Teeth Revisited: Proceedings of the VIIth International Symposium on Dental Morphology*, ed. D. E. Russell, J.-P. Santoro, and D. Sigogneau-Russell, pp. 133–143. *Mém. Mus. natn. Hist. nat., Paris, série C*, 53.

Pfretzschner, H.-U. (1992). Enamel microstructure and hypsodonty in large mammals. In *Structure, Function and Evolution of Teeth*, ed. P. Smith and E. Tchernov, pp. 147–162. London and Tel Aviv: Freund Publishing House Ltd.

Pfretzschner, H.-U. (1993). Enamel microstructure in the phylogeny of the Equidae. *J. Vert. Paleont.* 13:342–349.

Rasmussen, S. T., Patchin, R. E., Scott, D. B., and Heuer, A. H. (1976). Fracture properties of human enamel and dentin. *J. Dent. Res.* 55:154–164.

Rensberger, J. M. (1973). An occlusion model for mastication and dental wear in herbivorous mammals. *J. Paleont.* 47:515–528.

Rensberger, J. M. (1986). Early chewing mechanisms in mammalian herbivores. *Paleobiology* 12:474–494.
Rensberger, J. M. (1988). The transition from insectivory to herbivory in mammalian teeth. In *Teeth Revisited: Proceedings of the VIIth International Symposium on Dental Morphology*, ed. D. E. Russell, J.-P. Santoro, and D. Sigogneau-Russell, pp. 351–365. *Mém. Mus. natn. Hist. nat., Paris, série C*, 53.
Rensberger, J. M. (1992). Relationship of chewing stress and enamel microstructure in rhinocerotoid cheek teeth. In *Structure, Function and Evolution of Teeth*, ed. P. Smith and E. Tchernov, pp. 163–183. London and Tel Aviv: Freund Publishing House Ltd.
Rensberger, J. M. (1995). Determination of stresses in mammalian dental enamel and their relevance to the interpretation of feeding behaviors in extinct taxa. In *Functional Morphology in Vertebrate Paleontology*, ed. J. J. Thomason, pp. 151–172. Cambridge and New York: Cambridge University Press.
Rensberger, J. M. (1996). Differences in enamel fracture behavior and the evolution of dental strength in horses. *J. Vert. Paleont.* 16 (suppl. to 3):60A.
Rensberger, J. M. (1997). Mechanical adaptation in enamel. In *Tooth Enamel Microstructure*, ed. W. von Koenigswald and P. M. Sander, pp. 237–257. Rotterdam: Balkema.
Schaal, S., and Ziegler, W. (1993). *Messel – An Insight into the History of Life and of the Earth*. Oxford: Oxford University Press.
Sloan, R. E. (1987). Paleocene and latest Cretaceous mammal ages, biozones, magnetozones, rates of sedimentation, and evolution. *Geol. Soc. Amer. Special Paper* 209:165–200.
Sloan, R. E., and Van Valen, L. (1965). Cretaceous mammals from Montana. *Science* 148:220–227.
Waters, N. E. (1980). Some mechanical and physical properties of teeth. In *The Mechanical Properties of Biologic Materials*, ed. J. F. V. Vincent and J. D. Currey, pp. 95–135. *Symp. Soc. Exp. Biol. no.* 34. Cambridge and New York: Cambridge University Press.

CHRISTINE M. JANIS

7

Patterns in the evolution of herbivory in large terrestrial mammals: the Paleogene of North America

Introduction

This chapter examines the evolution of herbivory among the larger terrestrial mammals (ungulates and ungulate-like mammals) of the Early Tertiary (Paleogene, encompassing the Paleocene, Eocene, and Oligocene epochs). There are sound reasons for examining paleoecological trends in large mammals separately from small ones, both in terms of the allometry of the ecological attributes of living mammals (see Brown 1995) and in terms of paleontological and taphonomic biases (see Fortelius *et al.* 1996).

The term 'herbivory' is used to refer to diets that include primarily plant material, as opposed to animal material. In this chapter, I especially emphasize the evolution of the more restricted form of herbivory, folivory, meaning the consumption of only the fibrous, structural parts of the plants, such as stems and leaves. Folivorous mammals today include both browsers and grazers. (There were no grasslands in the Early Tertiary to support the latter dietary habit.) These mammals have some sort of symbiotic association with micro-organisms in the gut for cellulose fermentation. Folivores are usually contrasted with frugivores, that is animals that eat primarily fruit and other non-structural parts of the plant. There is also a category of 'frugivorous/folivorous' mammals, which take a mixture of fruit, seeds and leaves but do not select fibrous food requiring fermentation.

Exclusively herbivorous mammals must be above a certain body size (approximately one kilogram) because plant material does not contain enough nutrition per unit volume to support the metabolic rate of smaller endotherms (Kay 1975). Some small (i.e., less than 1 kg in body mass) extant mammals, such as voles and other specialized rodents, can be exclusively herbivorous by virtue of being 'granivorous,' specialized for taking grass seeds. Granivory may involve feeding on species of grasses, but in terms of the type of nutrition afforded the animal (i.e., feeding on the reproductive parts of plants rather than structural parts) it is more akin to frugivory than to grazing folivory (see Eisenberg 1981 for a discussion of mammalian feeding types).

The data in this chapter all come from the North American fossil record, although evolutionary trends in the higher latitudes of the Old World appear to be similar and are briefly discussed. Dietary preferences of extinct mammals are deduced from their dental features. In the case of the ungulates, at least, there is a diversity of modern taxa of known dietary habits so that the correlation between dental morphology and diet can be known with some assurance. The extinct ungulate-like mammals have no precise modern analogs, and thus their dental morphologies are more difficult to interpret precisely. However, as will be discussed, there are good biomechanical reasons for assigning particular diets to particular dental configurations. Patterns in the evolution of body size are also examined and discussed, since Paleogene mammals radiated into a wider variety of body sizes than were prevalent in the Mesozoic, when no mammal was bigger than a house cat. Ungulates have always been mostly of large body size compared with most other mammals, and the story of the evolution of herbivory in ungulates is in part the story of the evolution of larger body size.

The global climate and distribution of vegetation of the world was quite different in the Paleogene. Specifically, early Paleogene continents at temperate latitudes, such as North America, had much warmer mean annual temperatures (with concomitant lesser degrees of seasonality) than in the present day (Figure 7.1), with tropical-like forest habitats extending to within the confines of the Arctic Circle (Wolfe 1978, 1985). Following a peak in mean annual temperatures in northern latitudes at the early-middle Eocene boundary, temperatures started to decline, reaching a low in the early Oligocene. Concomitant with this climatic deterioration, tropical forests became more restricted towards the lower latitudes, and the major vegetational type in North America changed to that of a more seasonal deciduous type (summarized for North America in Wing 1998). The evolution of mammalian herbivory is examined against this background of climatic and vegetational changes.

This survey examines the generic richness through Paleogene times of ungulates and ungulate-like mammals distinguished both by dental

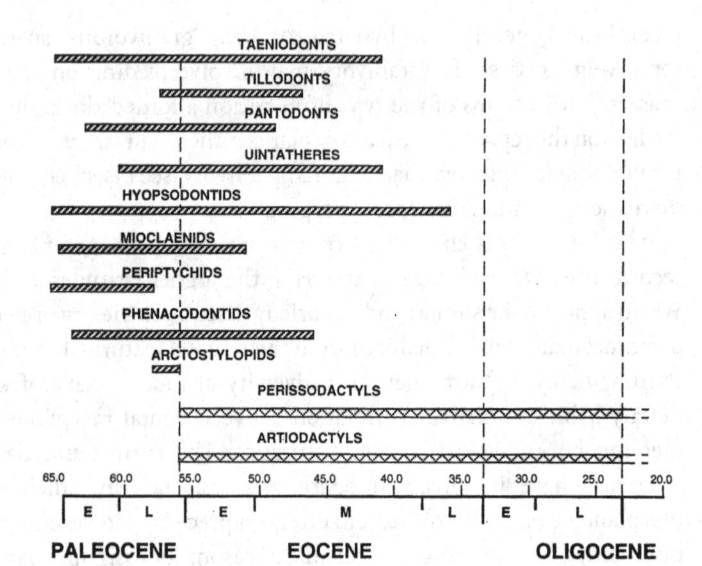

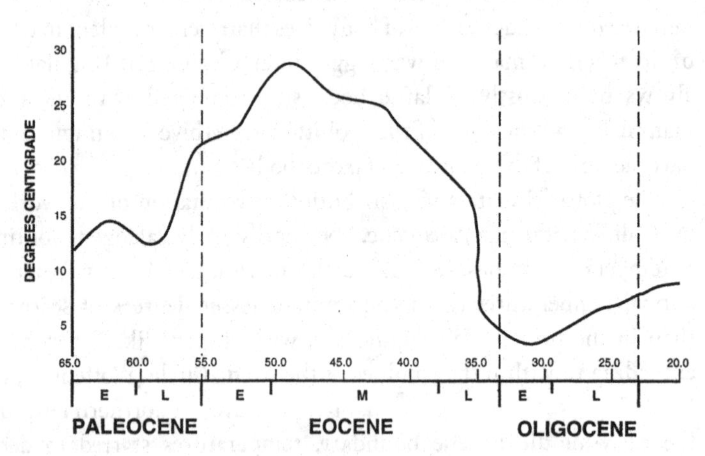

Figure 7.1. (A) Temporal ranges of lineages of North American ungulates and ungulate-like mammals. Key to shading: Cross-hatched lines, non-ungulates and archaic ungulates; diamond-patterned lines, modern ungulate orders. (B) Paleotemperatures in the North Sea for the Paleogene (modified from Burchardt 1978).

structure (= dietary type) and body size. Many of these general dietary trends have been discussed by other workers (e.g., Wing and Tiffney 1987; Collinson and Hooker 1991; Maas and Krause 1994; Gunnell *et al.* 1995). However, this chapter represents a new approach by attempting to quantify patterns of change in dental morphologies and also examining patterns in the evolution of body size, specifically focusing on ungulates and ungulate-like mammals. Jernvall, Hunter, and Fortelius (1996, 2000), convergently with the inception of this work, adopted a somewhat similar approach to the quantification of dental patterns, but their work has a broader geographic range (including all of the Northern Hemisphere) and uses faunal data at a coarser level of resolution. This chapter not only provides a quantification of some generally known trends but arrives at some novel conclusions, such as the speculation (based on the dietary habits of the ungulates) that the initial climatic event in the middle Eocene of the continental interior of the conterminous United States was drying rather than cooling.

Overview of the evolution of Early Tertiary herbivorous mammals

Although mammals have been a component of the terrestrial vertebrate fauna since the end of the Triassic, during the Mesozoic (the famed first two-thirds of their evolutionary history [Lillegraven, Kielan-Jaworowska, and Clemens 1979]) they were almost exclusively small (under 1 kg) and insectivorous or omnivorous. At this small body size they would have been unable to be exclusively herbivorous. Multituberculate mammals have been considered as Mesozoic herbivores, but despite their complex teeth, the dental structure and body size of most taxa would dictate a more omnivorous diet (Krause 1986). The recently discovered Cretaceous zhelestids, purported ungulate ancestors from the early Late Cretaceous of Uzbekistan (Archibald 1996), were likewise too small to have been exclusively herbivorous.

It was not until the Cenozoic, following the demise of the non-avian dinosaurs at the end of the Cretaceous, that mammals radiated into a larger variety of ecomorphological types, including larger body sizes. Thus the larger mammals in the Paleocene had the opportunity to include vegetation as a significant portion of their diet. However, although there was a great Paleocene diversification of taxa considered herbivorous, such as many lineages of 'condylarths,' or archaic ungulates, their dentitions and patterns of dental wear suggest that they were mainly omnivorous or generalist feeders (Janis 1979; Rensberger 1986; Hunter 1997). Some larger (up to bear-size during the Paleocene) ungulate-like taxa such as pantodonts and uintatheres had teeth apparently more adapted for shearing foliage; but the teeth of these animals were rather small in comparison with their bodies, quite unlike the 'batteries' of shearing teeth seen in modern folivorous ungulates, and had very simple shearing lophs (Janis 1979). Their teeth may have been adapted to masticate only soft, perhaps aquatic, types of vegetation. No analogs to these herbivores are known among modern mammals.

Terrestrial herbivores with ridged or lophed teeth adapted for a more extensive folivorous diet are not generally apparent until the Eocene, with some late Paleocene 'heralds' such as the more lophodont 'condylarth' *Meniscotherium* and the taxonomically enigmatic ungulate *Arctostylops*. This difference between Paleocene and Eocene forms in lineages generally regarded as 'herbivorous' is also apparent in the primates, in which there is also no evidence of folivorous taxa until the Eocene (Janis 1979; Collinson and Hooker 1991).

The start of the Eocene marked the first appearance of many modern mammalian orders, such as the perissodactyl and artiodactyl ungulates, and the true primates (Euprimates). (However, note that the earliest perissodactyls may be late Paleocene in age [Dashzeveg and Hooker 1997].) The appearance of these new groups did not immediately result in extinctions among the more archaic forms, at least in North America (Maas *et al.* 1995) (although early Eocene extinctions of archaic taxa were more marked in Europe [Russell 1975]). There was a rapid diversification of perissodactyls in the early Eocene; artiodactyl diversification occurred a little later, during the middle to late Eocene. Although many of the early members of these orders, such as the hyracotheriine horses and the dichobunid artiodactyls, had teeth suggestive of frugivory and selective browsing rather than true folivory (Janis 1979, 1990a; Collinson and Hooker 1991), by late Eocene times almost all perissodactyls and artiodactyls were larger animals, more clearly specialized for folivory, with the exception of the suiform (pig-like) artiodactyls that had divergently specialized for omnivory.

The Eocene thus appears to have been a critical time for the evolution of true herbivory in mammals. The early Eocene represented the earliest appearance of true folivores, and by the close of the epoch the diversity of trophic types of herbivorous mammals more closely resembled that seen in the modern fauna. The Eocene is also perhaps the most interesting Cenozoic epoch in terms of climatic changes, the late early to early middle Eocene representing the warmest time in the Cenozoic (see Figure 7.1), with the later Eocene temperatures of the higher latitudes plummeting in the transition between the 'Hot House' types of global environments of the Mesozoic and early Cenozoic and the 'Ice House' global environments of the later Cenozoic (Berggren and Prothero 1992 and references therein). This chapter attempts to explain some of the patterns in the diversity of ungulates, and of their dental features, during the Paleogene, and to relate them to changing global climates and the evolution of terrestrial herbivory.

Materials and methods

Taxonomic data

This chapter only examines North American mammals. The North American Paleogene record, especially in the Paleocene, is probably the best available in the world (Savage and Russell 1983). Certainly, the data are easier to obtain and to compile, due in part to the limited number of political boundaries in comparison with other continents. North America was isolated for much of the Paleogene, and due to its latitudinal position was strongly affected by the late Eocene climatic changes (see Janis 1993 and references therein). Similar, but not identical, patterns of faunal change among herbivorous mammals, especially ungulates, were seen on other northern continents (Jernvall *et al.* 1996, 2000), as will be discussed later.

The data on the number of taxa present in any one interval of time come from the contributions in Janis, Scott, and Jacobs (1998), specifically from chapters by Archibald, Cifelli and Schaff, Colbert and Schoch, Coombs, Effinger, Honey *et al.*, Kron and Manning, Lander, Lucas, Lucas and Schoch (a and b), Lucas, Schoch, and Williamson, MacFadden, Mader, Prothero (a–d), Stucky, Wall, Webb, and Wright (see references). Among the taxa of archaic ungulates ('condylarths') considered (data from Archibald 1998), clades with dentitions indicative of a more carnivorous/omnivorous diet, such as arctocyonids and mesonychids, were excluded from consideration. The taxonomic level of investigation was that of the genus; although species-level studies may be more appropriate for addressing some questions, the generic level may be considered as more appropriate for this broad-based paleoecological study (see, for example, Sepkoski and Kendrick 1993).

The Appendix to this chapter lists the taxa present in each subdivision of time, together with their dental morphology and body size category (see below). In a few instances, taxa were considered present in an interval if they were recorded from the time intervals preceding and succeeding that particular interval but not actually recorded from the interval itself.

Time intervals

The units of time selected for study follow the divisions of the North American land mammal 'ages' that were adopted in Woodburne (1987), and followed (with some updating revisions) in Janis *et al.* (1998). The land mammal 'ages' and their subdivisions are not of equal duration, nor are they equally well-sampled for fossil localities. As can be seen in Figures 7.3–7.6, units in the Paleocene may be less than a million years in length, whereas those in the later Eocene can be several million years long (for example, the entire Duchesnean Land Mammal age; it should be noted that the Duchesnean is a particularly poorly known interval of time [Lucas, 1992]). An attempt to investigate these sources of bias in these data is described elsewhere (Janis 1997–98). In summary, when the data are adjusted for the pattern of generic richness, there is only a slight dampening of the overall diversity pattern in select places, and, when time interval number is removed, the correlation between the number of localities and the number of genera is trivial. Possible effects of these biases are discussed later in the text.

Dental structure and dietary categories

The form of the cheek teeth of ungulates can be finely subdivided into many different detailed patterns (Jernvall 1995; Jernvall *et al.* 1996, 2000), but, for the purposes of this chapter, only three main types were considered (Figure 7.2). 'Bunodont' teeth represent the primitive quadritubercular form of tooth seen in ungulates, as well as in many other omnivorous mammals (Janis and Fortelius 1988). These are low-crowned, rounded-cusped teeth, present today in mammals with an omnivorous/frugivorous type of diet (e.g., bears, most pigs, and many primates including humans), that act to compress and pulp non-brittle, non-fibrous food in a pestle-and-mortar fashion (Lucas and Luke 1984).

The form of the cheek teeth in more herbivorous mammals shows an increase in the height of the low, isolated cusps, and a tendency to join them together to form high, cutting ridges, termed 'lophs' that act to shear and shred flat, tough, fibrous food such as leaves (Lucas and Luke

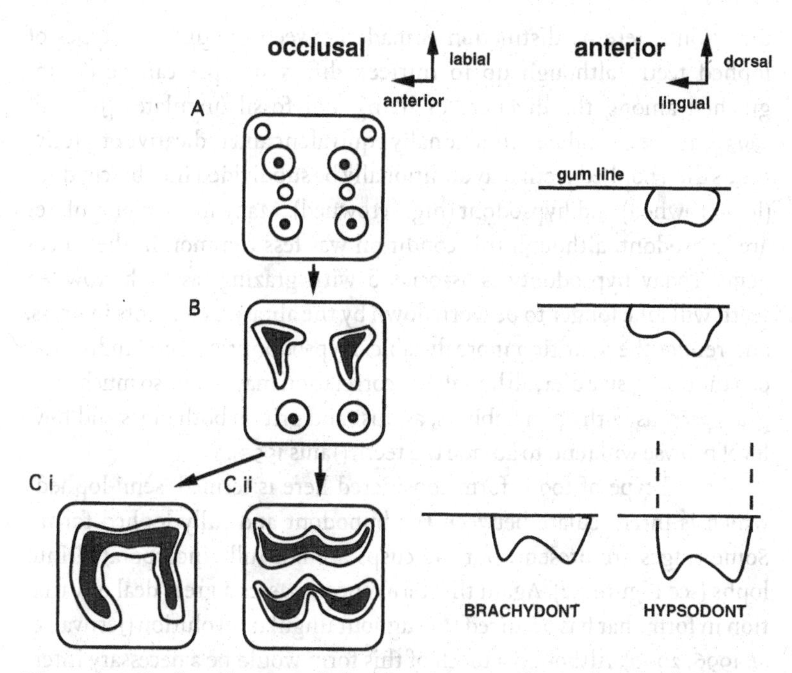

Figure 7.2. Cheek tooth patterns in herbivorous mammals, seen in occlusal and anterior views. Each example is represented by an upper left molar; shading indicates areas of exposed dentine (on a moderately worn tooth). (A) Bunodont tooth, indicative of an omnivorous/frugivorous type of diet. Examples in modern mammals include bush pig, bear, humans, and many other primates. (B) Semi-lophed tooth ('bunolophodont' or 'bunoselenodont'), indicative of a frugivorous/selective folivorous type of diet (i.e., non-fibrous parts of plant). Examples in modern mammals include mouse deer and tree hyraxes (plus leaf-eating primates like colobines). (C) Lophed tooth, indicative of a folivorous diet (i.e., fibrous parts of plant). Examples in modern animals include perissodactyls such as horses and rhinos (C i, 'lophodont') and ruminant artiodactyls such as deer and antelope (C ii, 'selenodont'). Lophed teeth can be low-crowned (brachydont, as in browsing mammals like deer and giraffe) or hypsodont (as in predominantly grazing mammals like horses and cows). Bunodont and semi-lophed teeth are almost invariably brachydont (with certain members of the extinct order Taeniodonta as the only known exception).

1984; Janis and Fortelius 1988). Such lophed teeth have developed from bunodont teeth on many occasions (Janis and Fortelius 1988; Janis 1995; Jernvall *et al.* 1996, 2000). Among the variety of possible patterns two basic ones are found among extant ungulates: the 'selenodont' teeth of ruminant artiodactyls (in which the lophs extend in a predominantly mesiodistal direction) and the 'lophodont' teeth of perissodactyls and rock hyraxes (in which the lophs extend in a predominantly labiolingual direction). Here no distinction is made between the different types of lophed teeth (although up to thirteen different types can be distinguished among the diversity of living and fossil ungulates [Jernvall 1995]); all are considered functionally equivalent and indicative of a folivorous diet. Lophed teeth may additionally be subdivided into brachydont (low-crowned) and hypsodont (high-crowned). Many modern ungulates are hypsodont, although this condition was less common in the Paleogene. Today hypsodonty is associated with grazing, as high-crowned teeth will take longer to be worn down by the abrasive elements in grass, and render the dentition more durable. Hypsodonty is a good indication of a more abrasive diet, although the correlation may not be so much with grass *per se* as with open habitats, as dust and grit on both grass and low-level browse will tend to abrade the teeth (Janis 1995).

A third type of tooth form considered here is termed 'semi-lophed', which is intermediate between the bunodont and fully lophed forms. Some ridges are present, but the cusps are not fully incorporated into lophs (see Figure 7.2). Again this term encompasses a great deal of variation in form that has occurred throughout ungulate evolution (Jernvall *et al.* 1996, 2000). Although a tooth of this form would be a necessary intermediate evolutionary stage in the acquisition of fully lophed teeth from an ancestral bunodont condition, it is also clearly adapted for a particular type of diet. Extant mammals that are selective frugivore/folivores, selecting predominantly the non-fibrous parts of the plant, have this form of cheek tooth (for example, the 'bunoselenodont' teeth of tragulids or the 'bunolophodont' teeth of tree hyraxes; Janis 1990a, 1995), and this type of tooth is also present in leaf-eating monkeys. Extinct mammals with this type of cheek tooth pattern are considered to have had a similar type of diet.

Very few extant terrestrial mammals have teeth that can be classified as semi-lophed. In the present-day African fauna, only three taxa fall into this category: the water chevrotain (*Hyemoschus aquaticus*), the tree hyrax (*Dendrohyrax dorsalis*), and the pygmy hippo (*Choeropsis liberensis*). Among the diversity of North American Paleogene ungulates, however, many of the archaic ('condylarth') families have members that could be considered semi-lophed, as do many primitive perissodactyls and artiodactyls. I have also included in this grouping mammals that do not fall within the clade Ungulata (Prothero, Manning, and Fischer 1988) but that nevertheless are 'ungulate-like' in their general structure and dentition; these include taeniodonts, tillodonts, pantodonts, and dinoceratans (uintatheres, which may or may not be true ungulates; see Lucas and Schoch 1998b). The Taeniodonta contain the only taxa that have teeth that may be considered 'semi-lophed hypsodont.' Their teeth have been classified as 'semi-lophed' for the purposes of this study, but actually the structure of their cheek teeth is quite different from those of ungulates and clearly independently derived.

It is a fundamental assumption of this chapter that dental patterns are evolutionarily plastic. The various types of lophed teeth have independently evolved from bunodont forms in numerous lineages, in marsupials as well as in placentals. I thus assume that the types of crown structure represent dental adaptations to the prevailing vegetational conditions, and do not merely reflect phylogenetic affinities. That is to say that a mammal possesses lophed cheek teeth because of its dietary habits, not because of its ancestry (although the precise form of the pattern of the lophs *will* most likely reflect its evolutionary history).

Body masses and size distributions

Body masses were estimated from dental measures (length of m2) provided in the chapters in Janis *et al.* (1988), which were then used in conjunction with the algorithms in Janis (1990b), or from estimates made directly by the authors of the contributions.

The body size categories were chosen to reflect natural groupings of mammals of different ecomorphological types. Although they fundamentally correspond to a logarithmic scale, the categories do not correspond to precise logarithmic means (such as 1–10 kg, 10–100 kg, etc.). There is no reason to assume that such numerical means represent biologically significant junctures. Additionally, as all ungulates are 'large' by mammalian standards (as the median mass of modern mammals ranges around one kilogram [Eisenberg 1981]) a traditional logarithmic scale would group the majority of mammals only into two size categories (i.e., 10–100 kg, and 100–1000 kg). The categories that have been chosen here appear to reflect natural ecomorphological 'breaks,' similar to those recognized by various other authors (Jarman 1974; Bertram and Biewener 1990; Fortelius, Van der Made, and Bernor 1996; Fortelius *et al.* 1996; Lewis 1997).

Size category 1 ('rabbit-sized') includes small terrestrial herbivores under 5 kg (among modern ungulates, encompassing small ruminants such as Asian tragulids and dwarf antelope such as dik-dik, and hyraxes). Size category 2 ('dog-sized') ranges from 5 kg to 25 kg (encompassing small antelope and deer such as duikers and roe deer, and peccaries). Size category 3 ('antelope-sized') ranges from 25 kg to 150 kg (including the majority of antelope and deer, plus llamas and pigs). Size category 4 ('horse-sized') ranges from 150 kg to 500 kg (including not only equids, but larger bovids and deer, such as cows and wapiti, the okapi, as well as tapirs and the pygmy hippo). Size category 5 ('rhino-sized') includes taxa over 500 kg (including not only rhinos, and other large herbivores such as hippos and elephants, but also very large artiodactyls such as giraffes and camels). This final size category might be better split into two if considering modern and Neogene mammals, separating the megaherbivores (over 1000 kg, or perhaps over 2000 kg) from the merely large ones; however, few Paleogene mammals (those under consideration here) attain weights in excess of 500 kg, and only a handful (some later Eocene uintatheres and brontotheres) would have weighed more than one metric ton.

Results

Tooth types

Figure 7.3 illustrates the generic diversity of ungulate and ungulate-like mammals with different tooth types over the duration of the Paleogene, and Figure 7.4 expresses these same data as percentage of the total fauna. The modern fauna, represented by the generic richness of dental types in present-day African ungulates (including subungulates), is shown to the right of the plot. Today there are few terrestrial herbivores with bunodont or semi-lophed teeth (although many arboreal primates have teeth of this type); the vast majority of African ungulates have lophed teeth, with about two-thirds of the genera having hypsodont teeth.

The modern situation contrasts rather sharply with the conditions in the Paleogene, especially during the early part of that time interval. At the start of the Paleocene, all of the ungulate taxa were bunodont; some early Paleocene condylarths actually had dental structures indicative of insectivory rather than of generalist omnivory (Hunter 1997). Even by the end of the Paleocene, bunodont taxa still comprised around 50% of the fauna. The slight fall in the numbers of bunodont taxa in the middle Paleocene, with a subsequent rise in the latest Paleocene and early Eocene, may reflect changing higher-latitude temperatures (see Figure 7.1). Semi-lophed taxa made their first appearance in the late early Paleocene, and increased in numbers through the epoch, but fully lophed taxa (e.g., arctostylopids) were not apparent until the very end of the Paleocene.

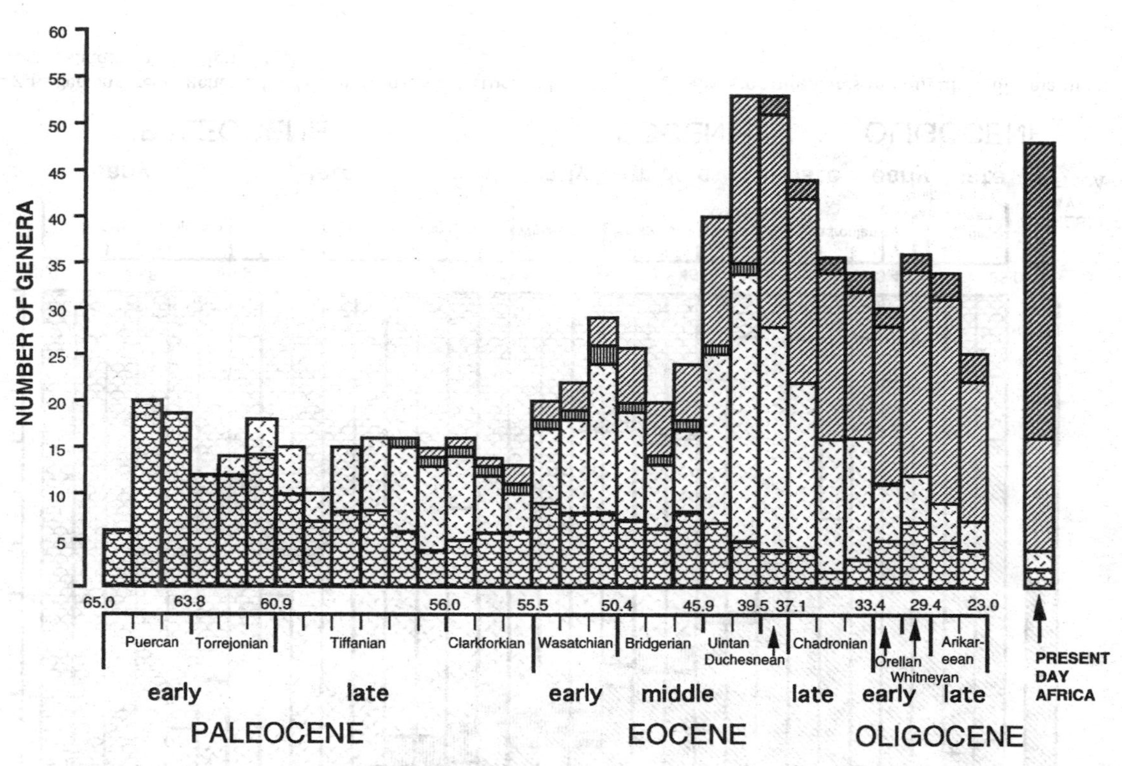

Figure 7.3. Generic diversity within dental structural types among Paleogene ungulates and ungulate-like mammals. Key to shading: tile or clam-shell pattern, bunodont; dashed pattern, semi-lophed (brachydont); vertical striping, semi-lophed (hypsodont); light cross-hatching, lophed (brachydont); heavy cross-hatching, lophed (hypsodont). Column on right-hand side shows pattern among present-day African ungulate genera for comparison.

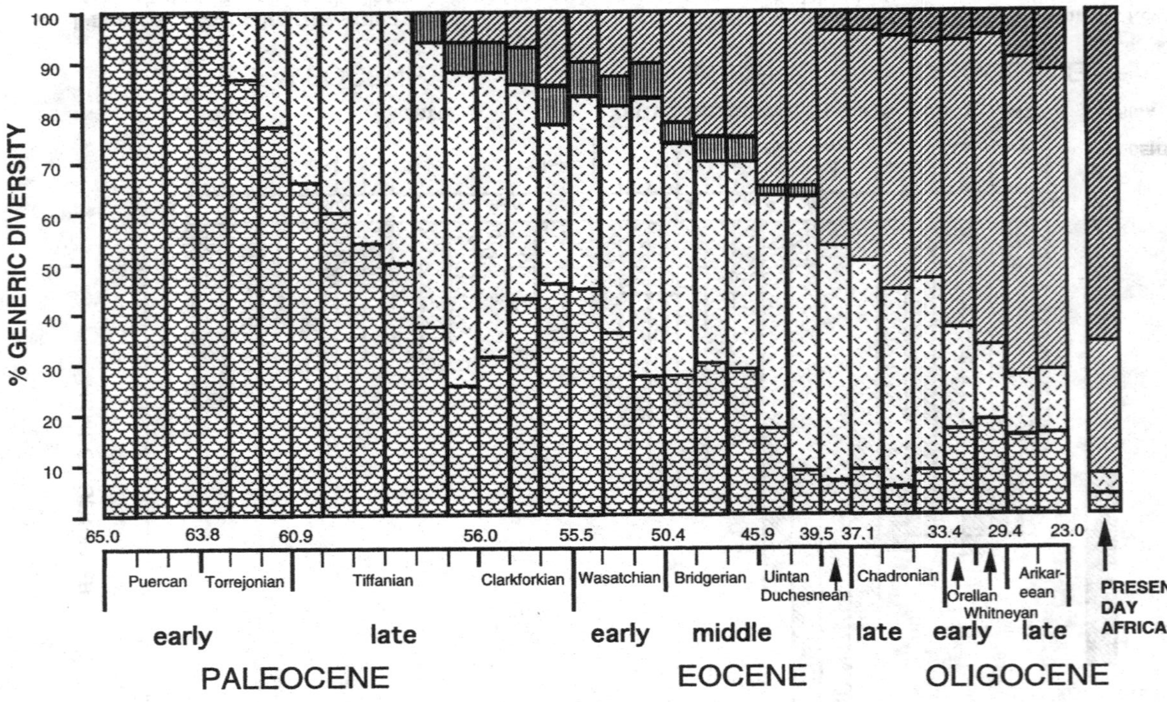

Figure 7.4. Percentages of generic diversity within dental structural types among Paleogene ungulates and ungulate-like mammals. Key to shading as for Figure 7.3.

Maas and Krause (1994) showed a category of 'herbivorous/folivorous' mammals (of all taxonomic categories, not just ungulates) existing at quite high generic diversities throughout the Paleocene, but their definitions of diets as defined by dental morphology must be different from the ones that I use here. I have defined folivorous taxa strictly by the possession of true dental lophs, whereas Maas and Krause (1994:108) defined these dietary types as having teeth with 'greater emphasis on shearing crests.' They did not state which taxa were actually included in this category.

Herbivores from the early and early middle Eocene (Wasatchian and Bridgerian Land Mammal ages) retained a similar diversity of tooth types to those of the latest Paleocene, with the numbers of lophed taxa hovering between 10% to 20% of the fauna, and the bunodont taxa maintaining their position at 40% to 50% of the fauna. The decrease in the percentage of bunodont taxa during the Wasatchian and into the early Bridgerian reflects an increase in the numbers of semi-lophed taxa rather than any real change in the numbers of bunodont taxa. Note that the Paleocene and early Eocene lophed taxa were predominantly of small body size, in contrast to the size of lophed ungulates in the modern fauna (see Figure 7.7). The early Eocene increase in generic diversity reflects the appearance of the modern ungulate orders Artiodactyla and Perissodactyla, in particular the radiation of equid and tapiroid perissodactyls.

A marked change occurred later in the middle Eocene, commencing in the late Bridgerian and peaking in the late Uintan and Duchesnean. The diversity of ungulate genera showed a significant increase, reflecting the diversification of ceratomorph (rhino-like) perissodactyls and of selenodont artiodactyls. Part of the magnitude of this increase may be due to the time-averaging effects in the longer intervals of the late Uintan and the Duchesnean, but there is clearly a change in the taxonomic diversity and composition of the ungulate fauna at this time. Most significantly, the numbers of bunodont taxa decreased dramatically, to around 10% of the herbivore fauna. This decrease represented the loss not only of archaic bunodont ungulates, such as 'condylarths,' but also the loss of the more primitive artiodactyls (Figure 7.7). However, the difference in the faunal percentages was not made up at this time by the expansion of taxa with lophed teeth, as might be expected from the fact that lophed ungulates form the predominant dental type in the modern fauna. Rather, it was by the expansion of semi-lophed taxa, which comprised around 50% of the large herbivore fauna for the remainder of the epoch. The percentages of lophed taxa increased slightly towards the end of the Eocene, with the first hypsodont taxa appearing at the end of the middle Eocene, but lophed taxa did not come to comprise more than 50% of the fauna until the Oligocene. Note also that generic diversity of ungulates decreased from a peak in the late middle Eocene to late Oligocene levels approaching those of the early Eocene (Figure 7.3).

By the end of the Oligocene, the diversity of tooth types in the ungulate fauna approached the modern condition, with the exception of the low numbers of hypsodont taxa. The rise in the number of bunodont taxa in the Oligocene reflects the radiation of the larger, more derived suiform artiodactyls, such as anthracotheres, entelodonts, and peccaries (see Figure 7.7).

Body sizes

The distribution of body sizes through the Paleogene reveals a rather different type of information about diversity patterns (Figures 7.5–7.6). Among present-day African ungulates, all sizes are fairly equally represented; the smallest number of genera are the 'dog-sized' ones in category 2, and the largest numbers are in the 'antelope-sized' category 3. This, again, is very different from the situation during the early Paleogene. At the start of the Paleocene, all the terrestrial herbivores were small, bunodont condylarths in size categories 1 and 2 (see also Figure 7.7). Later in the early Paleocene larger forms appeared, to comprise around 50% of the herbivorous fauna by the late Paleocene. These larger taxa were the ungulate-like mammals: taeniodonts, tillodonts, pantodonts and dinoceratans (Figure 7.7). Interestingly enough, these animals almost all fall into size categories 4 and 5, so that the Paleocene herbivore communities were made up of small, bunodont animals and large, semi-lophed ones, with almost no taxa of medium size. Note, however, that these larger taxa were absolutely rare as faunal components (R. Stucky, pers. comm.). The single genus representing size category 3 is also the one hypsodont semi-lophed taeniodont.

By the early Eocene, the number and proportion of smaller herbivores increased, a reflection of the appearance of the perissodactyls and artiodactyls. No very large herbivores were present, reflecting the extinction of the barylambdid pantodonts (the hippo-like Eocene pantodont, *Coryphodon*, was not as large as the more ground-sloth-like Paleocene barylambdids). The number of taxa in size category 3 increased slightly, reflecting the evolution of larger lophed tapiroids (see Figure 7.7). By the middle

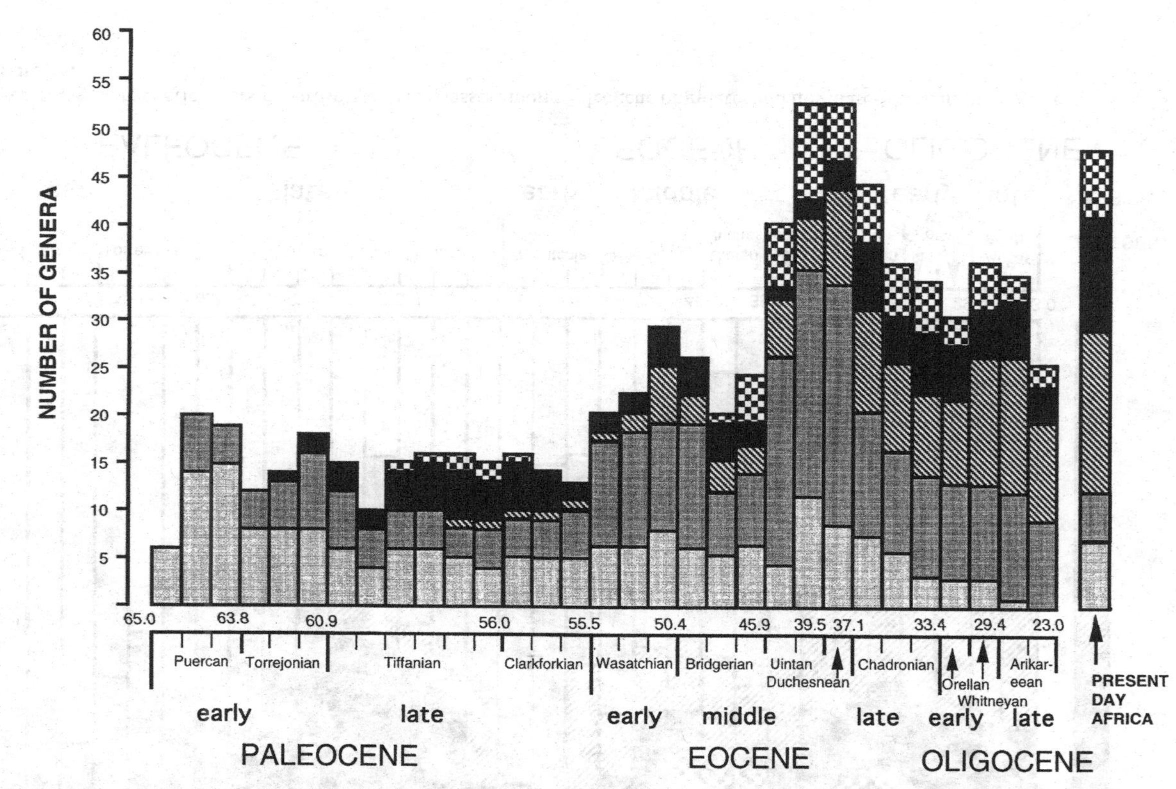

Figure 7.5. Generic diversity within body size classes among Paleogene ungulates and ungulate-like mammals. Key to shading: lightest tone, Size 1 ('rabbit-sized', <5 kg); medium-light tone, Size 2 ('dog-sized', 5–25 kg); cross-hatching, Size 3 ('antelope-sized', 25–150 kg); medium-dark tone, Size 4 ('horse-sized', 150–500 kg); checkered tone, Size 5 ('rhino-sized', >500 kg). Column on right-hand side shows present-day African ungulate genera for comparison.

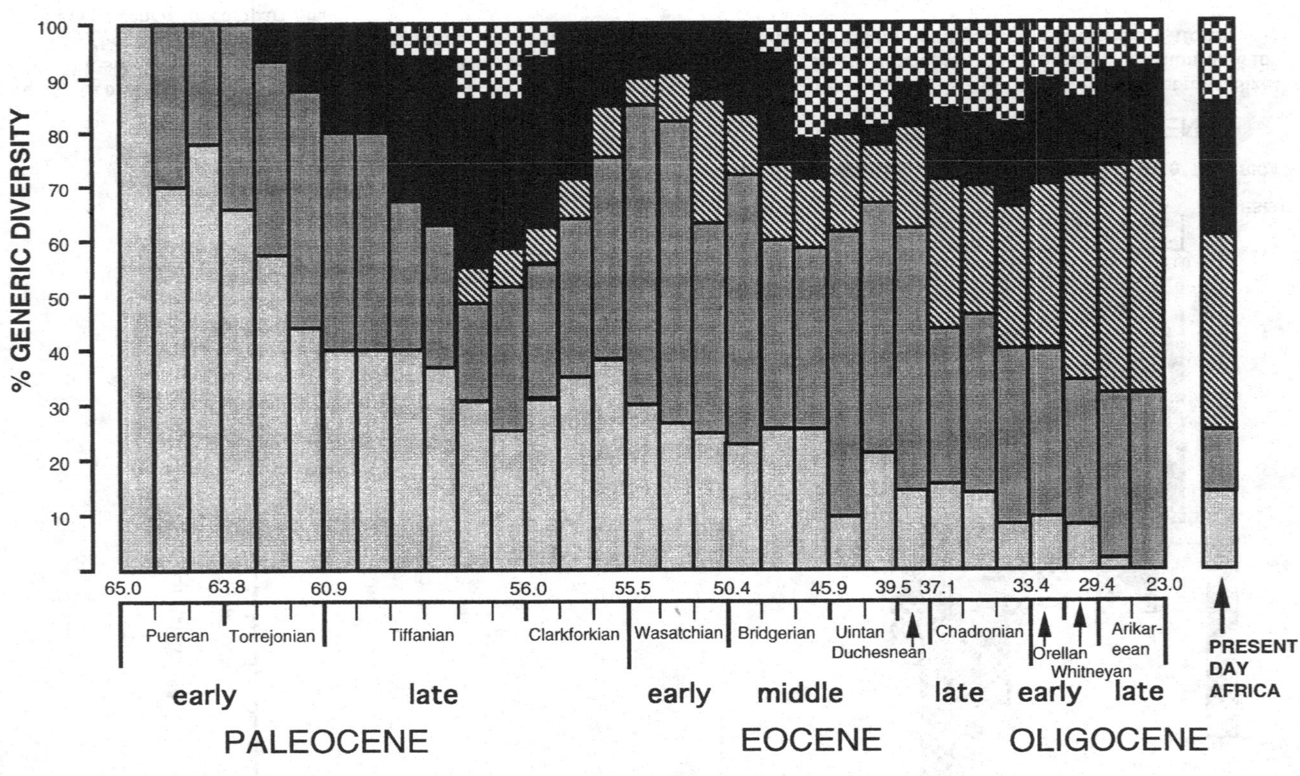

Figure 7.6. Percentages of generic diversity within body size classes among Paleogene ungulates and ungulate-like mammals. Key to shading as for Figure 7.5.

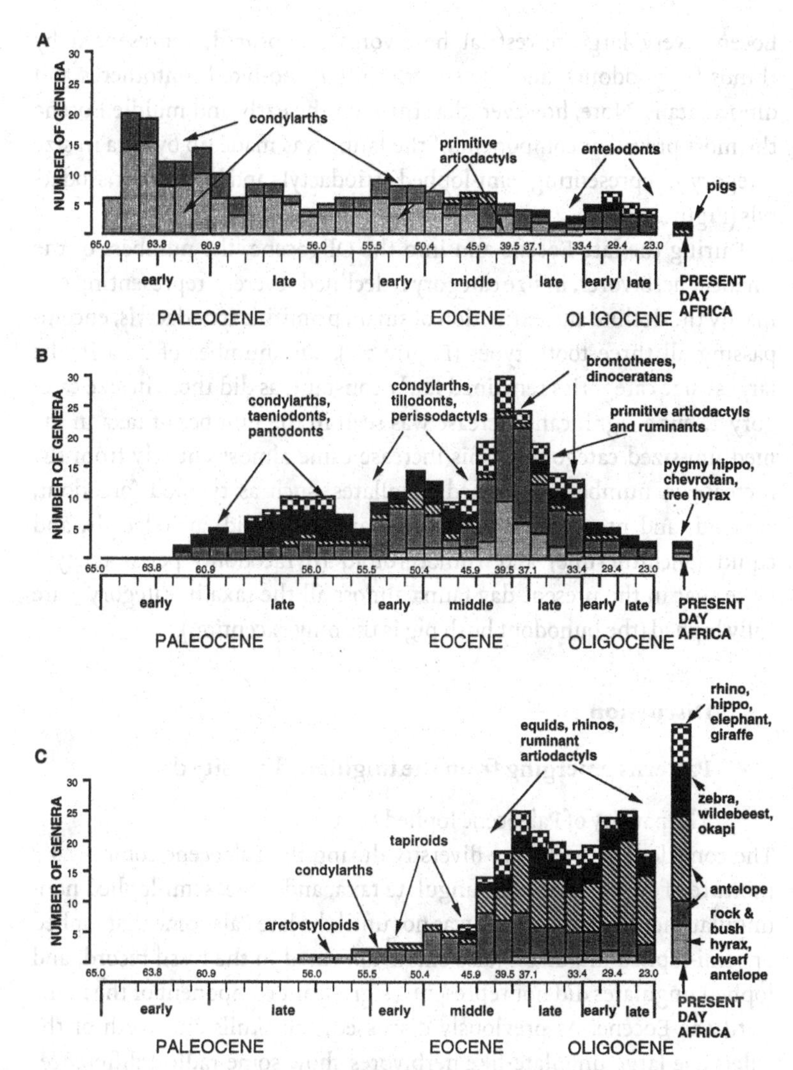

Figure 7.7. Generic diversity within different body size classes, broken down by dental structural type. (A) Bunodont taxa. (B) Semi-lophed taxa. (C) Lophed taxa. Key to shading as for Figure 7.5.

Eocene, very large terrestrial herbivores reappeared, represented by rhinos (amynodonts) and the superficially rhino-like brontotheres and dinoceratans. Note, however, that through the early and middle Eocene the most prevalent component of the fauna was made up by taxa in size category 2, representing semi-lophed artiodactyls and lophed perissodactyls (Figure 7.7).

During the late Eocene and into the Oligocene, the numbers of the smallest herbivores, in size category 1, declined severely, representing primarily the decline and extinction of small, primitive artiodactyls, encompassing all three tooth types (Figure 7.7). The number of taxa in the largest size categories remained fairly constant, as did those in size category 2, but a significant increase was seen in the number of taxa in the medium-sized category 3. This increase came almost entirely from the increase in numbers of lophed ungulates, such as tylopod (oreodont, camelid, and protoceratid) and ruminant (traguloid) artiodactyls, and equid (anchitheriine) and rhinocerotoid (hyracodont) perissodactyls. Note that in the present-day fauna almost all the taxa in category 3 are fully lophed (the bunodont bush pig is the only exception).

Discussion

Patterns emerging from the ungulate diversity data

The paucity of Paleocene lophed taxa

The considerable herbivore diversity during the Paleocene comprised a mixture of small, bunodont ungulate taxa, and large, semi-lophed non-ungulate herbivorous taxa. It was not until the late Paleocene that lophed or semi-lophed ungulates (also small) appeared in the fossil record, and lophed ungulates did not represent a significant component of the fauna until the Eocene. As previously discussed, the skulls and teeth of the Paleocene large ungulate-like herbivores show some radical differences from those of extant large herbivores, with an apparently small tooth volume for their body size (Janis 1979). It thus seems unlikely that these animals had the type of folivorous diet (fibrous, with a large volume of intake) found in present-day large folivorous ungulates. It appears more likely that the type of terrestrial folivory encountered today was not prevalent until the Eocene.

Almost all the Paleogene bunodont ungulates were small, or small-to-intermediate in size (i.e., under 25 kg). Extant bunodont ungulates (pigs) are mainly medium-sized or of medium-large size, as was characteristic for the North American fauna by the Oligocene. Similarly, the lophed ungulate taxa were initially of fairly small size: it was not until the late middle Eocene that larger lophed taxa (rhinocerotoids) first emerged. This suggests that the warmer world of the early Paleogene supported a larger diversity of smallish terrestrial herbivores at temperate latitudes than found in modern faunas at any latitude, perhaps indicative of a very different climatic and vegetational regime from anywhere in the world today (see discussion below).

The diversity increase of the late middle Eocene

The general pattern of generic diversity in the northern latitudes rather closely follows the paleotemperature curve (Figure 7.1) for mammals (Stucky 1990, 1992) and for amphibians and reptiles (Hutchison 1982). There was an increase in both diversity and temperature over the Paleocene/Eocene boundary, with a peak in the late early to early middle Eocene, followed by a decline to the end of the Paleogene. Gingerich (1984) especially noted how the diversity of primates in the northern latitudes closely followed the paleotemperature curve for the entire Cenozoic. The faunal changes of the middle and late Eocene of North America were marked by the extinction of many archaic Eocene taxa, such as the archaic ungulates ('condylarths') and ungulate-like taxa considered here, and also among other taxa such as plesiadapiforms (early ?primates), true primates (adapiforms and omomyids), archaic carnivorous mammals ('miacoid' carnivorans and creodonts), and archaic insectivorous mammals ('proteutherians'), as well as non-eutherian taxa such as multituberculates and didelphoid marsupials (Savage and Russell 1983). While many archaic mammals became extinct, many of the more modern groups of mammals (for example, the modern families of carnivorans and artiodactyls) made their first appearance in the late Eocene. But the overall pattern was a decline in diversity.

However, the diversity of large terrestrial herbivores did not closely follow this overall trend. Diversity increased through the Paleocene/Eocene boundary, and up to the middle Eocene, followed by an initial decline in the Bridgerian (early middle Eocene), as with the majority of other mammals. However, there was a marked, further increase in diversity in the later middle Eocene, peaking in the late Uintan and Duchesnean, and then declining back to early Eocene values in the late Eocene (Chadronian). The absolute magnitude of this increase (a doubling of generic diversity between early middle and late middle Eocene) may be inflated; for example, many late middle Eocene primitive artiodactyl taxa are known from a single or very few occurrences (see Stucky 1998). However, the primary pattern of increase is real, even when taxa known from restricted occurrences are eliminated (Janis 1997–98). A peak in generic diversity of ungulates in the late Eocene (rather than the late middle Eocene), is reported by Jernvall *et al.* (1996, 2000), for Europe and in Asia, as well as in North America. (However, note that the data in this chapter are more up-to-date than those used by Jernvall *et al.*, which were derived from Savage and Russell 1983 [J. Hunter, pers. comm].)

In North America, the late middle to late Eocene (Uintan–Chadronian) marked the appearance of many new taxa, such as tapirid and rhinocerotoid perissodactyls, and families of suiform and selenodont (tylopod plus ruminant) artiodactyls. The majority of these new taxa probably represent an immigration event from Asia at this time (see Janis 1993). Thus, regardless of the nature of the events during the later Eocene that were detrimental to archaic mammals, and to arboreal modern mammals such as primates in the higher latitudes, they were favorable to the diversification of terrestrial herbivores.

The late middle Eocene increase in semi-lophed taxa

One might imagine that the supposed cooling and drying of the later Eocene (see Berggren and Prothero 1992 and references therein) would represent a situation advantageous to modern types of ungulates – mainly large (size categories 3–5), folivorous animals with hypsodont lophed teeth, living in a cool and dry world today. However, the ungulate taxa of this type did not form the initial diversification at the start of the 'Ice House' world in the late middle Eocene. Rather the bulk of the diversity of Uintan and Duchesnean taxa was made up of small (under 25 kg), semi-lophed, primitive ('dichobunids') and early selenodont artiodactyls (camelids and traguloids), and small to medium-sized (10–150 kg) equid and tapiroid perissodactyls, almost none of them hypsodont. Jernvall *et al.* (2000) show a similar late Eocene rise (and subsequent Oligocene decline) in their 'one-lophed' taxa (broadly corresponding to my category of 'semi-lophed').

There is no equivalent to this type of diversity of terrestrial herbivorous taxa in present-day tropical forests (see also Andrews 1992). Perhaps the closest equivalent today would be in the forests of tropical Central and South America, where there is a diversity of 'ungulate-like' caviomorph rodents such as agoutis and pacas. In contrast to Old World tropical forests, where the only small terrestrial ungulates are tragulids (sparse in both taxonomic diversity and absolute numbers), the New World tropical forests have a greater abundance of these large rodents, which have been perceived as playing the role of an ungulate-like ecomorph (Dubost 1968). However, these present-day tropical forests also contain a large diversity of arboreal herbivores, and thus cannot be precisely compared with the late Eocene of North America when the numbers of arboreal herbivores, such as primates, were in rapid decline.

The types, and body size diversity, of lophed taxa held fairly steady from the late Eocene through to the end of the Oligocene. But there was a large change in the number and size diversity of semi-lophed taxa over this time period. The large taxa, the brontotheres and dinoceratans (size category 5), were all extinct by the end of the Eocene. Among the smaller ones, the 'dichobunid' artiodactyls (mostly size category 1) almost all went extinct (with the exception of the leptochoerids), and the more derived selenodont ones, such as ruminants and camelids (mostly size category 2), were replaced by related, fully lophodont forms. By the end of the Oligocene, the taxonomic diversity of semi-lophed taxa in North America was as reduced as it is in present-day Africa, and the remaining taxa occupied only the two smallest size categories. Jernvall *et al.* (2000) note a reappearance of a small number of taxa with a single loph in the late Oligocene and Miocene (apparently represented by the re-immigration of chalicotheres, now representing animals in size category 4 rather than category 2, although I would probably term these taxa as 'fully lophed').

Another interesting point noted by Jernvall *et al.* (1996, 2000) is the difference in the patterns of diversity and disparity in ungulate dental types between the Paleogene and the Neogene. Whereas ungulate generic diversity rose again in the Miocene, following the fall in the late Paleogene, and the *disparity* of dental types remained high (both bunodont forms and forms with four or more dental lophs are apparent), the *diversity* of tooth forms never reached the level seen in the Paleogene. Only seven crown types (out of a total range of 28) are present among extant ungulates, whereas, during the later Eocene, up to a dozen or so different crown types were present at any one time interval. Jernvall *et al.* interpret these data as indicating a greater number of environmental opportunities for ungulates in later Eocene habitats than is found today, although it is not clear how many of these 'niches' have been taken over by rodents and other mammals in today's ecosystems. The pattern of dental diversity among present-day small herbivores, such as rodents and lagomorphs, really needs to be included in order to complete this picture. Rodent diversity and abundance greatly increased in the late middle Eocene in North America, with the appearance of forms with complexly lophed cheek teeth (eomyids and zapodids), and, furthermore, lagomorphs made their first appearance at this time (Webb 1989). In Europe, there was also a diversification of rodents with teeth indicative of a folivorous diet (theridomyids) during the late Eocene (Collinson and Hooker 1991). It is unlikely that small rodents or lagomorphs would have been in direct competition with larger ungulates for the same resources; rather, this increase in rodent diversity most likely reflects a climatic shift, as discussed later.

The interaction between size categories and dental types

Among present-day ungulates, the majority of African taxa are in the middle size range (category 3), and the majority also have lophed teeth. However, it is perhaps surprising to see that, among Paleogene terrestrial herbivores, size category 3 was almost completely occupied by lophed herbivores (Figure 7.7). Paleocene faunas, almost completely lacking in lophed taxa (and the true lophed taxa at the end of the epoch being only small animals), contained small (bunodont) and large (semi-lophed) terrestrial herbivores, but no medium-sized ones. In fact, the rise of the medium-sized terrestrial herbivores largely reflects the rise to predominance of forms with lophed teeth. The biological significance of this correlation of body size with dental structure (and hence with diet) is not clear, as the animals in the modern fauna clearly demonstrate that it is *possible* to be a lophed folivore at a range of body sizes. It must then be the case that more omnivorous and selective browsing taxa are somehow *avoiding* this size category (as hinted at by the diversity of modern African animals, but the sample size is small). Again, the biological reasons for this are unclear.

Evolutionary and ecological implications

As discussed above, these data suggest that full folivory, as indicated by the presence of taxa with fully lophed teeth, was not seen in North American mammalian communities until the latest Paleocene, and that a radiation of more diverse, larger-bodied folivores did not occur until the late middle Eocene. Other studies of Paleocene mammals reveal little evidence of folivorous forms (Wing and Tiffney 1987; Stucky 1990; Krause and Maas, 1990; Collinson and Hooker 1991). The reasons for this must lie in paleoenvironmental factors. The ten million years of the Paleocene certainly represented a sufficient amount of time for the evolution of a wide variety of mammalian sizes and body types. There is no reason to assume that this time period would have been insufficient for the evolution of adaptations for folivory. One speculation concerning early Cenozoic ecosystems is that the demise of the dinosaurs would have resulted in a significant change in environmental conditions (Wing and Tiffney 1987; Stucky 1990). The large herbivorous dinosaurs of the Late Cretaceous may have browsed the herbage to such levels that an open type of habitat was created and sustained, much as elephants maintain certain types of open habitats today. Sudden removal of the large herbivores may have resulted in a floral 'rebound,' creating dense forest with herbage that would be inaccessible to potential small mammalian herbivores. Alternatively (or perhaps additionally), a more equable climate in the Paleocene in comparison with that of the Late Cretaceous may have favored a closed canopy woodland over the evidently more open habitats of the late Mesozoic (Wolfe and Upchurch 1986; Wolfe 1990). The bolide impact at the Cretaceous/Tertiary boundary may also have had a long-term impact on the North American vegetation, resulting in low floristic diversity (Wing 1998), which may have had an effect on the diversity of terrestrial herbivores. Large seeds in the Paleocene evidently provided food for terrestrial omnivores (Wing and Tiffney 1987), but the leaves may have been too fibrous (see Janis 1989).

Given the above speculation, what changes could have caused a restructuring of the habitat in the Eocene, such that the vegetation was now accessible to small (under 25 kg) terrestrial ungulates (as demonstrated by their lophed or semi-lophed teeth)? Warming during the late Paleocene may have made the vegetational habitat more favorable for small ungulates; the increase in mean annual temperature and decrease in temperature seasonality at this time should have resulted in a less seasonal supply of food resources, as there would have been less annual fluctuation in leaf and fruit production (Wing 1998). The effects of this warming event may explain the first appearance of fully lophed taxa at this time. The faunas of the higher latitudes in the early Eocene provide evidence of a tropical forest-like habitat, with a high diversity of arboreal primates and gliders, but the terrestrial forms were much more diverse than in the Paleocene (Collinson and Hooker 1987; Rose 1990; Stucky 1990; Andrews 1992). However, the combination of mammalian ecomorphs, with a high diversity of both terrestrial frugivores and arboreal insectivores, has no precise counterpart in any present-day habitat (Collinson and Hooker 1991; Andrews 1992).

The diversity of lophed and semi-lophed ungulates in the early Eocene of North America suggests a different type of forest habitat. There must have been a more open forest floor, with a less dense tree canopy, and more understory growth to provide food for these animals. This inferred greater spacing of the trees suggests that the higher latitudes during the early Eocene were not only warmer but also drier than in the Paleocene, or at least with more seasonally pulsed episodes of rainfall. The apparent increased palatability of herbage to the mammals also suggests a more seasonal type of vegetation, where the leaves would be less fibrous and hence more easily digested (Deinum and Dirven 1975). Gunnell *et al.* (1995), in an analysis of the entire North American faunal community, also concluded that the early Eocene may have been drier than the Paleocene, and Wolfe (1987) reported a predominance of deciduous plants occurring with the warming that occurred in North America from the late Paleocene into the early Eocene.

Although the 'heyday' of the evolution of large herbivorous mammals is often seen as being correlated with the spread of the savanna grasslands in the Miocene (Webb 1977, 1989; MacFadden, this volume), reflected by a peak in generic diversity (slightly exceeding late Eocene levels) on all three northern continents, it should be appreciated that much of the morphological diversity of ungulate tooth crown types present in the late Eocene was lost by the end of the Paleogene (Jernvall *et al.* 1996, 2000). The late Eocene represented a peak time for the diversification of small-to-medium-sized ungulates, types of animals that are rare components of the modern, mainly large-bodied, ungulate fauna.

Why the later part of the Eocene, which was the time of extinction of so many of the archaic mammals, should favor the diversification of these smaller, semi-lophed ungulates remains a bit of a mystery. Both paleobotanical and faunal diversity data indicate an increase in seasonality during the later Eocene. The vegetation in higher latitudes was less tropical in nature in the late middle Eocene than earlier in the epoch (Wolfe 1985; Collinson and Hooker 1987). Increasing seasonal dryness affected the vegetation in the continental interior (Wing 1998), and, in the late Eocene, tropical vegetation was replaced by more temperate types, such as mixed coniferous and deciduous woodlands, and winter frosts may have been present (Wolfe 1978, 1985; Collinson, Fowler and Boulter 1981). An increased abundance of small seeds in the late Eocene is suggestive of seasonal plant growth cycles (Wing and Tiffney 1987). A more open, less forested, type of habitat for the late Eocene at higher latitudes is also suggested by changes in the postcranial structure of rodents and ungulates, with the appearance of taxa with cursorial adaptations, and the extinction of arboreal mammals such as many primates (Webb 1978, 1989; Stucky 1990, 1992). Changes in diversity patterns are seen not only among the high-latitude mammals, but also among the amphibians and reptiles, which declined in diversity in the later Eocene (Hutchison 1982).

Increase in seasonality may have made more normally fibrous parts of plants more readily available, as such environmental conditions result in greater differentiation in fiber content between the leaves and the stem (Deinum and Dirvan 1975). This vegetational change would have especially affected small, selective-feeding artiodactyls, which formed a large component of the late Eocene diversification (see Janis 1989). However, the later Eocene has been traditionally interpreted as a time of climatic cooling (Berggren and Prothero 1992 and references therein). While the disappearance of most frugivores, and of obligate arboreal mammals such as primates, coincides with this scenario, it is difficult to envisage a cool, seasonal climate containing a diversity of tragulid-like, selective-browsing artiodactyls, ungulates that today are confined to tropical environments. Yet the semi-lophed 'dichobunids', tylopods, and ruminants that diversified the late middle Eocene appear to have been almost identical in craniodental structure to the extant tragulids. The abrupt drop in diversity of semi-lophed taxa in the Oligocene corresponds with a dramatic temperature fall in the earliest Oligocene, when mean annual temperatures in North America declined by about 13 °C in as little as one million years (Prothero and Heaton 1996; Wing 1998).

A possible window on the conditions of the late middle Eocene is provided by the work of Badgley and Fox (in review), who documented climatic variables and ecological structure for modern mammalian communities in all of North America. Canonical ordination of climatic and ecological variables resulted in localities being ordinated in multidimensional space, with a primary/first axis representing seasonality of temperature (high seasonality corresponding to cold climates, low seasonality corresponding to low climates of today), and a secondary axis representing moisture stress (high precipitation versus high evapotranspiration). The ecological structure (size structure and trophic structures) of faunas was plotted in the same multidimensional space. Figure

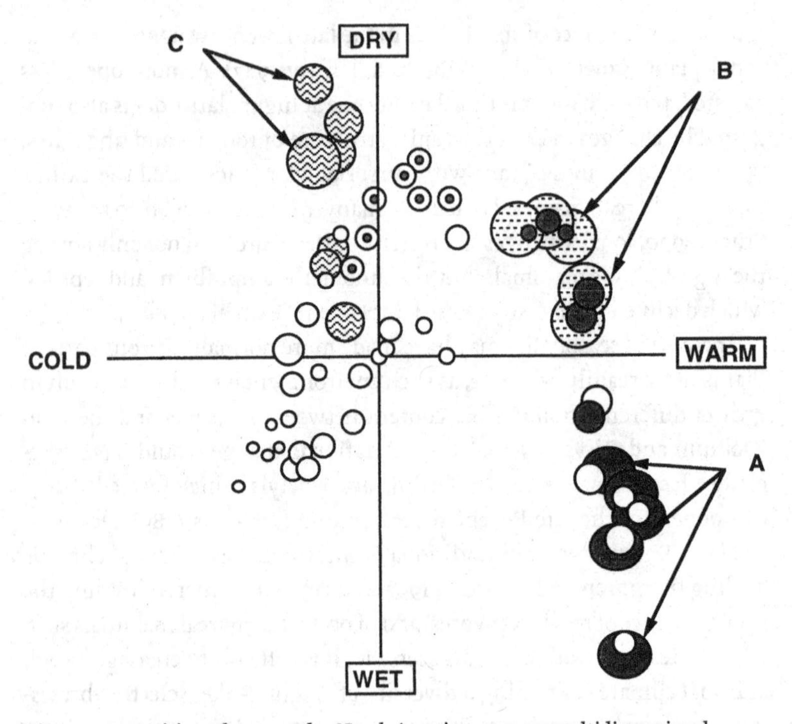

Figure 7.8. Position of present-day North American taxa on multidimensional environmental axes (from Badgley and Fox, in review), with possible positions of North American Paleogene taxa superimposed. Circles indicate where particular modern localities fall in the multidimensional climate space, and the size of the circle indicates the numbers of modern species in that locality. Key to the shading of the circles: filled circles, frugivorous taxa; unfilled circles, herbivorous and granivorous taxa (dotted circles, preponderance of small, granivorous taxa; wavy line circles, preponderance of large, folivorous taxa). (A) probable placement in multidimensional climate space of Paleocene–early Eocene bunodont ungulates and primates. (B) probable placement in multidimensional climate space of middle-late Eocene ungulates (semi-lophed). (C) probable placement in multidimensional climate space of Oligocene and Neogene folivorous ungulates (lophed).

7.8 represents a compilation of several of their figures, and includes a 'best guess' estimate of where Paleogene mammals would fall in this multidimensional space. Note that present-day frugivorous taxa (small to medium-sized and represented by primates, rodents, bats and procyonid carnivores) fall mainly in the warm/wet quadrant, herbivorous taxa (including large ungulates and small rodents) fall mainly in the cold/dry quadrant, whereas granivorous taxa (mainly of small size) fall mainly in the warm/dry quadrant.

Although it is impossible to make a precise extrapolation from the modern-day taxa and climates to those of the North American Paleogene, the following interpretation seems likely. Paleogene primates and other herbivorous mammals probably had similar environmental demands to present-day primates, and would have occupied the warm-wet quadrant of the multidimensional space (Figure 7.8, 'A'). Likewise the lophed, folivorous herbivorous ungulates of the Oligocene would probably have occupied the cold-dry quadrant of the space, as do the ungulates today (Figure 7.8, 'C'). The major speculation I advance is that the diversity of small, or small to medium-sized semi-lophed ungulate taxa of the late middle Eocene, for which there is no direct modern analog, would be broadly analogous to the 'granivorous' taxa of Badgley and Fox (today largely represented by rodents), falling in the warm-dry quadrant in the multidimensional space (Figure 7.8, 'B').

While the dental structure of the later Eocene small semi-lophed artiodactyls resembles that of tragulids, which today inhabit moist, tropical habitats, it is not too dissimilar (in terms of functional equivalence) to that of bunolophodont granivorous rodents. That is, these teeth could equally well be used to masticate seeds in a more arid environment as well as buds in a more moist one. A change from a tropical, non-seasonal climate to a colder, seasonal one (as has always been hypothesized for the later Eocene) would affect selective feeders equally in both terrestrial and arboreal environments. In contrast, a climatic change predominantly in the direction of aridity might have significant effects on arboreal mammals by affecting the structure of the forest canopy and the seasonality of large fruit production, but have a lesser effect on the availability of seed and pod resources derived from lower-level plants for terrestrial selective herbivores.

Therefore, the composition of the late middle Eocene ungulate fauna, with the increased diversity of the semi-lophed taxa that are most likely indicative of warm-dry conditions (or that at least contradict the notion of a cool climate), does not suggest that cooling was a significant issue until the late Eocene (Chadronian) and, especially, during the Oligocene. Although paleobotanical data also suggest a drying trend in the later Eocene of North America (Frederiksen 1995; Wing 1998), there has been, as far as I am aware, no other proposal that middle Eocene drying occurred in the absence of significant cooling (although Frederiksen [1995] did suggest that drying may be the main event). This suggestion of an initial middle Eocene interval of drying without cooling is at variance with the paleotemperature data obtained from the benthic foraminifera, but may better reflect Paleogene conditions in the continental interior. Alternatively, it may be the case that cooling had an insignificant effect on the flora and fauna until such a point that a significant frost season was invoked (C. Badgley, pers. comm.).

This chapter has been concerned with the faunal diversities and probable paleoecological conditions in North America. Despite broad similarities in the Eocene across the northern continents in ungulate taxonomic and morphological diversity (Jernvall *et al.* 1996, 2000), details of floral and faunal evolution may well have been different on different continents. For example, angiosperm pollen shows that floral diversity is greatest during the middle middle Eocene in North America, but in the early early Eocene in western Europe; differing paleogeographic configurations and rates of species origination and extinction may account for these differences (Frederiksen 1995). Note, however, that Wing, Alroy, and Hickey (1995) did not support these conclusions for North America; their data, based on whole plant fossils, show a maximum diversity in the earliest Eocene (although these differences between the researchers may relate to the different areas of North America under study). J. J. Hooker (pers. comm.) notes that, among ungulate tooth types, taxa with bilophodont teeth (like those of modern tapirs) were more common in western Europe during the Eocene than in North America. This difference may merely represent the phylogenetic composition of the two areas (the diversity of ceratomorph perissodactyls was considerably different, with tapiroids being more common in Europe and rhinocerotoids more common in North America), or it may represent a difference in dental types that reflected different environmental conditions. The data in this chapter have generated some interesting speculations about Paleogene changes in paleoecology and paleoclimates in North America; it remains to be seen if these patterns hold up globally, or in North America with more detailed study.

Acknowledgments

I thank Hans-Dieter Sues and Conrad Labandeira for inviting me to participate in their symposium 'Origin and Evolution of Terrestrial Herbivory' at the Sixth North American Paleontological Convention (1996), and Catherine Badgley, Robert Emry, Jeremy Hooker, John Hunter, and Richard Stucky for comments on the original manuscript.

References

Andrews, P. (1992). Community evolution in forest habitats. *J. Hum. Evol.* 18:326–343.

Archibald, J. D. (1996). Fossil evidence for a Late Cretaceous origin of 'hoofed' mammals. *Science* 272:1150–1153.

Archibald, J. D. (1998). Archaic ungulates ('Condylarthra'). In *The Evolution of Tertiary Mammals of North America*, ed. C. M. Janis, K. M. Scott, and L. L. Jacobs, pp. 292–331. Cambridge and New York: Cambridge University Press.

Badgley, C., and Fox, J. L. In review. Ecological structure of North American mammalian faunas in relation to climate. *Journal of Biogeography*.

Berggren, W. A., and Prothero, D. R. (1992). Eocene–Oligocene climatic and biotic evolution: an overview. In *Eocene–Oligocene Climatic and Biotic Evolution*, ed. D. R. Prothero and W. A. Berggren, pp. 1–28. Princeton: Princeton University Press.

Bertram, J. E. A., and Biewener, A. A. (1990). Differential scaling of the long bones in the terrestrial Carnivora and other animals. *J. Morphol.* 20:157–169.

Brown, J. H. (1995). *Macroecology*. Chicago: University of Chicago Press.

Burchardt, B. (1978). Oxygen isotope paleotemperatures from the Tertiary period in the North Sea area. *Nature* 275:121–123.

Cifelli, R. L., and Schaff, C. R. (1998). Arctostylopida. In *The Evolution of Tertiary Mammals of North America*, ed. C. M. Janis, K. M. Scott, and L. L. Jacobs, pp. 332–336. Cambridge and New York: Cambridge University Press.

Colbert, M. W., and Schoch, R. M. (1998). Tapiroidea and other moropomorphs. In *The Evolution of Tertiary Mammals of North America*, ed. C. M. Janis, K. M. Scott, and L. L. Jacobs, pp. 569–582. Cambridge and New York: Cambridge University Press.

Collinson, M. E., Fowler, M. K., and Boulter, M. C. (1981). Floristic changes indicate a cooling climate in the Eocene of southern England. *Nature* 291:315–317.

Collinson, M. E. and Hooker, J. J. (1987). Vegetational and mammalian fauna changes in the early Tertiary of southern England. In *The Origin of Angiosperms and their Biological Consequences*, ed. E. M. Friis and W. G. Chaloner, pp. 259–303. Cambridge and New York: Cambridge University Press.

Collinson, M. E., and Hooker, J. J. (1991). Fossil evidence of interactions between plants and plant-eating mammals. *Phil. Trans. R. Soc. Lond.* B 333:197–208.

Coombs, M. C. (1998). Chalicotherioidea. In *The Evolution of Tertiary Mammals of North America*, ed. C. M. Janis, K. M. Scott, and L. L. Jacobs, pp. 560 – 568. Cambridge and New York: Cambridge University Press.

Dashzeveg, D., and Hooker, J. J. (1997). New ceratomorph perissodactyls (Mammalia) from the Middle and Late Eocene of Mongolia: their implications for phylogeny and dating. *Zool. J. Linn. Soc.* 120:105–138.

Deinum, B., and Dirven, J. G. P. (1975). Climate, nitrogen and grass. 6. Comparison of yield and chemical composition of some temperate and tropical grass species grown at different temperatures. *Neth. J. Agric. Sci.* 23:69–82.

Dubost, G. (1968). Les niches écologiques de forêts tropicales sud-américaines et africaines, sources de convergences remarquables entre Rongeurs et Artiodactyles. *La Terre et la Vie* 22:3–28.

Effinger, J. A. (1998). Entelodontidae. In *The Evolution of Tertiary Mammals of North America*, ed. C. M. Janis, K. M. Scott, and L. L. Jacobs, pp. 375–380. Cambridge and New York: Cambridge University Press.

Eisenberg, J. F. (1981). *The Mammalian Radiations*. Chicago: University of Chicago Press.

Fortelius, M., Van der Made, J., and Bernor, R. L. (1996). Middle and late Miocene Suoidea of central Europe and the eastern Mediterranean: evolution, biogeography, and paleoecology. In *The Evolution of Western Eurasian Neogene Mammal Faunas*, ed. R. Bernor, V. Fahlbusch, and H.-W. Mittmann, pp. 348–377. New York: Columbia University Press.

Fortelius, M., Werdelin, L., Andrews, P., Bernor, R. L., Gentry, A., Humphrey, L., Mittman, H.-W., and Viranta, S. (1996). Provinciality, diversity, turnover, and paleoecology in land mammal faunas of the later Miocene of Western Eurasia. In *The Evolution of Western Eurasian Neogene Mammal Faunas*, ed. R. Bernor, V. Fahlbusch, and H.-W. Mittmann, pp. 414–448. New York: Columbia University Press.

Frederiksen, N. O. (1995). Differing Eocene floral histories in southeastern North America and western Europe: influence of paleogeography. *Hist. Biol.* 10:13–23.

Gingerich, P. D. (1984). Primate evolution. In *Mammals: Notes for a Short Course*, ed. P. D. Gingerich and C. E. Badgley, pp. 167–181. Knoxville: University of Tennessee, Department of Geological Sciences, Studies in Geology, 8.

Gunnell, G. F., Morgan, M. E., Maas, M. C., and Gingerich, P. D. (1995). Comparative paleoecology of Paleogene and Neogene mammalian faunas: trophic structure and composition. *Palaeogeogr., Palaeoclimatol., Palaeoecol.* 115:265–286.

Honey, J. G., Harrison, J. A., Prothero, D. R., and Stevens, M. S. (1998). Camelidae. In *The Evolution of Tertiary Mammals of North America*, ed. C. M. Janis, K. M. Scott, and L. L. Jacobs, pp. 439–462. Cambridge and New York: Cambridge University Press.

Hunter, J. P. (1997). *Adaptive Radiation of Early Paleocene Ungulates*. Unpublished Ph.D. dissertation, State University of New York at Stony Brook.

Hutchison, J. H. (1982). Turtles, crocodilians, and champsosaur diversity changes in the Cenozoic of the north-central region of Western United States. *Palaeogeogr., Palaeoclimatol., Palaeoecol.* 37:149–151.

Janis, C. M. (1979). *Aspects of the Evolution of Ungulate Mammals*. Unpublished Ph.D. dissertation, Harvard University. Cambridge, MA.

Janis, C. M. (1989). A climatic explanation for patterns of evolutionary diversity in ungulate mammals. *Palaeontology* 32:463–481.

Janis, C. M. (1990a). The correlation between diet and dental wear in herbivorous mammals, and its relationship to the determination of diets of extinct species. In *Evolutionary Paleobiology of Behavior and Coevolution*, ed. A. J. Boucot, pp. 241–259. Amsterdam and New York: Elsevier.

Janis, C. M. (1990b). Correlation of cranial and dental variables with body size in ungulates and macropodoids. In *Body Size in Mammalian Paleobiology: Estimation and Biological Implications*, ed. J. Damuth and B. J. MacFadden, pp. 255–300. Cambridge and New York: Cambridge University Press.

Janis, C. M. (1993). Tertiary mammal evolution in the context of changing climates, vegetation, and tectonic events. *Annu. Rev. Ecol. Syst.* 24: 467–500.

Janis, C. M. (1995). Correlation between craniodental morphology and feeding behavior in ungulates: reciprocal illumination between living and fossil taxa. In *Functional Morphology in Vertebrate Paleontology*, ed. J. J. Thomason, pp. 76–98. Cambridge and New York: Cambridge University Press.

Janis, C. M. (1997–98). Ungulate teeth, diets, and climatic changes at the Eocene/Oligocene boundary. *Zoology* 100:203–220.

Janis, C. M., and Fortelius, M. (1988). On the means whereby mammals achieve increased functional durability of their dentitions, with special reference to limiting factors. *Biol. Rev.* 63:197–230.

Janis, C. M., Scott, K. M., and Jacobs, L. L. (eds.) (1998). *Evolution of Tertiary Mammals of North America*. Cambridge and New York: Cambridge University Press.

Jarman, P. J. (1974). The social organisation of antelopes in relation to their ecology. *Behaviour* 48: 213–267.

Jernvall, J. (1995). Mammalian molar cusp patterns: Developmental mechanisms of diversity. *Acta Zool. Fenn.* 198:1–61.

Jernvall, J., Hunter, J. P., and Fortelius, M. (1996). Molar tooth diversity, disparity, and ecology in Cenozoic radiations. *Science* 274:1489–1492.

Jernvall, J., Hunter, J. P., and Fortelius, M. (2000). Trends in the evolution of molar crown types in ungulate mammals: evidence from the Northern Hemisphere. In *Function and Evolution of Teeth*, ed. M. Teaford, M. Ferguson, and M. M. Smith, pp. 269–281. Cambridge and New York: Cambridge University Press.

Kay, R. F. (1975). The functional adaptations of primate molar teeth. *Am. J. Phys. Anthropol.* 43:195–216.

Krause, D. W. (1986). Competitive exclusion and taxonomic displacement in the fossil record: the case of rodents and multituberculates in North America. *Contrib. Geol. Univ. Wyoming, Spec. Pap.* 3:93–117.

Krause, D. W., and Maas, M. C. (1990). The biogeographic origins of late Paleocene–early Eocene mammalian immigrations to the western interior of North America. In *Dawn of the Age of Mammals in the Northern Part of the Rocky Mountains*, ed. T. M. Bown and K. D. Rose. *Geol. Soc. Am. Spec. Pap.* 243:71–105.

Kron, D. G., and Manning, E. (1998). Anthracotheriidae. In *The Evolution of Tertiary Mammals of North America*, ed. C. M. Janis, K. M. Scott, and L. L. Jacobs, pp. 381–388. Cambridge and New York: Cambridge University Press.

Lander, B. (1998). Oreodontoidea. In *The Evolution of Tertiary Mammals of North America*, ed. C. M. Janis, K. M. Scott, and L. L. Jacobs, pp. 402–425. Cambridge and New York: Cambridge University Press.

Lewis, M. E. (1997). Carnivoran paleoguilds of Africa: implications for hominid food procurement strategies. *J. Hum. Evol.* 32: 257–288.

Lillegraven, J. A., Kielan-Jaworowska, Z., and Clemens, W. A. (eds.) (1979). *Mesozoic Mammals: The First Two-Thirds of Mammalian History*. Berkeley: University of California Press.

Lucas, P. W., and Luke, D. A. (1984). Chewing it over: basic principles of food breakdown. In *Food Acquisition and Processing in Primates*, ed. D. J. Chivers, B. A. Wood, and A. Bilsborough, pp. 283–302. New York: Plenum Press.

Lucas, S. G. (1992). Redefinition of the Duchesnean land mammal 'age', late Eocene of western North America. In *Eocene–Oligocene Climatic and Biotic Evolution*, ed. D. R. Prothero and W. A. Berggren, pp. 88–105. Princeton: Princeton University Press.

Lucas, S. G. (1998). Tillodontia. In *The Evolution of Tertiary Mammals of North America*, ed. C. M. Janis, K. M. Scott, and L. L. Jacobs, pp. 268–278. Cambridge and New York: Cambridge University Press.

Lucas, S. G., and Schoch, R. M. (1998a). Pantodonta. In *The Evolution of Tertiary Mammals of North America*, ed. C. M. Janis, K. M. Scott, and L. L. Jacobs, pp. 274–283. Cambridge and New York: Cambridge University Press.

Lucas, S. G., and Schoch, R. M. (1998b). Dinocerata. In *The Evolution of Tertiary Mammals of North America*, ed. C. M. Janis, K. M. Scott, and L. L. Jacobs, pp. 284–291. Cambridge and New York: Cambridge University Press.

Lucas, S. G., Schoch, R. M., and Williamson, T. E. (1998). Taeniodonta. In *The Evolution of Tertiary Mammals of North America*, ed. C. M. Janis, K. M. Scott, and L. L. Jacobs, pp. 260–267. Cambridge and New York: Cambridge University Press.

Maas, M. C., and Krause, D. W. (1994). Mammalian turnover and community structure in the Paleocene of North America. *Hist. Biol.* 8:91–128.

Maas, M. C., Anthony, M. R. L., Gingerich, P. D., Gunnell, G. F., and Krause, D. W. (1995). Mammalian generic diversity and turnover in the Late Paleocene and Early Eocene of the Bighorn and Crazy Mountain Basins, Wyoming and Montana (USA). *Palaeogeogr., Palaeoclimatol., Palaeoecol.* 115:181–207.

MacFadden, B. J. (1998). Equidae. In *The Evolution of Tertiary Mammals of North America*, ed. C. M. Janis, K. M. Scott, and L. L. Jacobs, pp. 537–559. Cambridge and New York: Cambridge University Press.

Mader, B. J. (1998). Brontotheriidae. In *The Evolution of Tertiary Mammals of North America*, ed. C. M. Janis, K. M. Scott, and L. L. Jacobs, pp. 525–536. Cambridge and New York: Cambridge University Press.

Prothero, D. R. (1998a). Oromerycidae. In *The Evolution of Tertiary Mammals of North America*, ed. C. M. Janis, K. M. Scott, and L. L. Jacobs, pp. 426–430. Cambridge and New York: Cambridge University Press.

Prothero, D. R. (1998b). Protoceratidae. In *The Evolution of Tertiary Mammals of North America*, ed. C. M. Janis, K. M. Scott, and L. L. Jacobs, pp. 431–438. Cambridge and New York: Cambridge University Press.

Prothero, D. R. (1998c). Hyracodontidae. In *The Evolution of Tertiary Mammals of North America*, ed. C. M. Janis, K. M. Scott, and L. L. Jacobs, pp. 589–594. Cambridge and New York: Cambridge University Press.

Prothero, D. R. (1998d). Rhinocerotidae. In *The Evolution of Tertiary Mammals of North America*, ed. C. M. Janis, K. M. Scott, and L. L. Jacobs, pp. 595–605. Cambridge and New York: Cambridge University Press.

Prothero, D. R., and Heaton, T. H. (1996). Faunal stability during the Early Oligocene climatic crash. *Palaeogeogr., Palaeoclimatol., Palaeoecol.* 127:257–283.

Prothero, D. R., Manning, E., and Fischer, M. (1988). The phylogeny of the ungulates. In *The Phylogeny and Classification of Tetrapods, Vol.* 2, ed. M. J. Benton, pp. 201–234. *Systematics Association Special Volume* No. 35B. Oxford: Clarendon Press.

Rensberger, J. M. (1986). Early chewing mechanisms in mammalian herbivores. *Paleobiology* 47:474–494.

Rose, K. D. (1990). Postcranial skeletal remains and adaptations in early Eocene mammals from the Willwood Formation, Bighorn Basin, Wyoming. In *Dawn of the Age of Mammals in the Northern Part of the Rocky Mountains*, ed. T. M. Bown and K. D. Rose. *Geol. Soc. Am. Spec. Pap.* 243:107–113.

Russell, D. E. (1975). Paleoecology of the Paleocene–Eocene transition in Europe. In *Approaches to Primate Paleobiology*, ed. F. S. Szalay, pp. 28–61. Basel: Karger.

Savage, D. E., and Russell, D. E. (1983). *Mammalian Paleofaunas of the World*. Reading, MA: Addison-Wesley Publishing Company.

Sepkoski, J. J., Jr., and Kendrick, D. C. (1993). Numerical experiments with model monophyletic and paraphyletic taxa. *Paleobiology* 19:168–184.

Stucky, R. K. (1990). Evolution of land mammal diversity in North America during the Cenozoic. In *Current Mammalogy*, ed. H. H. Genoways, pp. 375–432. New York: Plenum Press.

Stucky, R. K. (1992). Mammalian faunas in North America of Bridgerian to early Arikareean 'ages' (Eocene and Oligocene). In *Eocene–Oligocene Climatic and Biotic Evolution*, ed. D. R. Prothero and W. A. Berggren, pp. 464–493. Princeton: Princeton University Press.

Stucky, R. K. (1998). Eocene bunodont and bunoselenodont Artiodactyla ('dichobunids'). In *The Evolution of Tertiary Mammals of North America*, ed. C. M. Janis, K. M. Scott, and L. L. Jacobs, pp. 358–374. Cambridge and New York: Cambridge University Press.

Wall, W. P. (1998). Amynodontidae. In *The Evolution of Tertiary Mammals of North America*, ed. C. M. Janis, K. M. Scott, and L. L. Jacobs, pp. 583–588. Cambridge and New York: Cambridge University Press.

Webb, S. D. (1977). A history of savanna vertebrates in the New World. Part I: North America. *Annu. Rev. Ecol. Syst.* 8:355–380.

Webb, S. D. (1989). The fourth dimension in North American terrestrial mammalian communities. In *Patterns in the Structure of Mammalian Communities*, ed. D. W. Morris, Z. Abramsky and M. R. Willig, pp. 181–203. *Special Publication of the Museum*, No. 28. Lubbock: Texas Tech University.

Webb, S. D. (1998). Hornless ruminants. In *The Evolution of Tertiary Mammals of North America*, ed. C. M. Janis, K. M. Scott, and L. L. Jacobs, pp. 463–476. Cambridge and New York: Cambridge University Press.

Wing, S. L. (1998). Tertiary vegetation of North America as a context for mammalian evolution. In *The Evolution of Tertiary Mammals of North America*, ed. C. M. Janis, K. M. Scott, and L. L. Jacobs, pp. 37–65. Cambridge and New York: Cambridge University Press.

Wing, S. L., and Tiffney, B. H. (1987). The reciprocal interaction of angiosperm evolution and tetrapod herbivory. *Rev. Palaeobot. Palynol.* 50:179–210.

Wing, S. L., Alroy, J., and Hickey, L. J. (1995). Plant and mammal diversity in the Paleocene to Early Eocene of the Bighorn Basin. *Palaeogeogr., Palaeoclimatol., Palaeoecol.* 115:117–155.

Wolfe, J. A. (1978). A paleobotanical interpretation of Tertiary climates in the Northern Hemisphere. *Am. Sci.* 66:694–703.

Wolfe, J. A. (1985). Distribution of major vegetational types during the Tertiary. *Geophys. Monogr.* 32:357–375.

Wolfe, J. A. (1987). Late Cretaceous–Cenozoic history of deciduousness and the terminal Cretaceous event. *Paleobiology* 13:215–226.

Wolfe, J. A. (1990). Palaeobotanical evidence for a marked temperature increase following the Cretaceous/Tertiary boundary. *Nature* 343:153–156.

Wolfe, J. A., and Upchurch, G. R. (1987). North American nonmarine climates and vegetation during the Late Cretaceous. *Palaeogeogr., Palaeoclimatol., Palaeoecol.* 61:33–77.

Woodburne, M. O. (ed.) (1987). *Cenozoic Mammals of North America*. Berkeley: University of California Press.

Wright, D. B. (1998). Tayassuidae. In *The Evolution of Tertiary Mammals of North America*, ed. C. M. Janis, K. M. Scott, and L. L. Jacobs, pp. 389–401. Cambridge and New York: Cambridge University Press.

Appendix

Database for herbivorous mammals from the Paleogene of North America considered in this chapter, arranged by stratigraphic provenance. Columns (from left to right) represent order, family, genus, dental type and size class (see text for explanation). Notes: Underlined dental type, hypsodont; n.g., unnamed new genus; *inc.sed.*, *incertae sedis*.

PALEOCENE

Puercan 1

'Condylarthra'	Hyopsodontidae	*Litomylus*	Bunodont	1
		Oxyprimus	Bunodont	1
	Periptychidae	*Anisonchus*	Bunodont	1
		Conacodon	Bunodont	1
		Maiorana	Bunodont	1
		Mimatuta	Bunodont	1

Puercan 2

Taeniodonta		*Onychodectes*	Bunodont	2
		Schochia	Bunodont	2
		Wortmania	Bunodont	2
'Condylarthra'	Hyopsodontidae	*Litomylus*	Bunodont	1
	Mioclaenidae	*Bubogonia*	Bunodont	2
		Choeroclaenus	Bunodont	1
		Ellipsodon	Bunodont	1
		Litaletes	Bunodont	1
		Promioclaenus	Bunodont	1
		Protoselene	Bunodont	1
		Tiznatzinia	Bunodont	1
	Periptychidae	*Anisonchus*	Bunodont	1
		Carsioptychus	Bunodont	2
		Conacodon	Bunodont	1
		Ectoconus	Bunodont	2
		Escatepos	Bunodont	1
		Gillisonchus	Bunodont	1
		Haploconus	Bunodont	1
		Hemithlaeus	Bunodont	1
		Oxyacodon	Bunodont	1

Puercan 3

Taeniodonta		*Onychodectes*	Bunodont	2
		Wortmania	Bunodont	2

PALEOCENE (*cont.*)

Puercan 3 (*cont.*)

'Condylarthra'	Hyopsodontidae	*Haplaletes*	Bunodont	1
		Litomylus	Bunodont	1
	Mioclaenidae	*Choeroclaenus*	Bunodont	1
		Ellipsodon	Bunodont	1
		Litaletes	Bunodont	1
		Promioclaenus	Bunodont	1
		Protoselene	Bunodont	1
		Tiznatzinia	Bunodont	1
	Periptychidae	*Anisonchus*	Bunodont	1
		Carsioptychus	Bunodont	2
		Conacodon	Bunodont	1
		Ectoconus	Bunodont	2
		Gillisonchus	Bunodont	1
		Haploconus	Bunodont	1
		Hemithlaeus	Bunodont	1
		Oxyacodon	Bunodont	1
		Tinuviel	Bunodont	1

Torrejonian 1

Taeniodonta		*Conoryctella*	Bunodont	2
'Condylarthra'	Hyopsodontidae	*Haplaletes*	Bunodont	1
		Litomylus	Bunodont	1
	Mioclaenidae	*Ellipsodon*	Bunodont	1
		Litaletes	Bunodont	1
		Mioclaenus	Bunodont	2
		Promioclaenus	Bunodont	1
		Protoselene	Bunodont	1
	Periptychidae	*Anisonchus*	Bunodont	1
		Haploconus	Bunodont	1
		Periptychus	Bunodont	2
	Phenacodontidae	*Tetraclaenodon*	Bunodont	2

Torrejonian 2

Taeniodonta		*Conoryctella*	Bunodont	2
Pantodonta		*Pantolambda*	Semi-lophed	4
'Condylarthra'	Hyopsodontidae	*Haplaletes*	Bunodont	1
		Litomylus	Bunodont	1
	Mioclaenidae	*Ellipsodon*	Bunodont	1
		Litaletes	Bunodont	1
		Mioclaenus	Bunodont	2
		Promioclaenus	Bunodont	1

PALEOCENE (*cont.*)

Torrejonian 2 (*cont.*)				
		Protoselene	Bunodont	1
	Periptychidae	*Anisonchus*	Bunodont	1
		Haploconus	Bunodont	1
		Periptychus	Bunodont	2
	Phenacodontidae	*Tetraclaenodon*	Bunodont	2
Torrejonian 3				
Taeniodonta		*Conoryctes*	Bunodont	2
		Huerfanodon	Bunodont	2
		Psittacotherium	Semi-lophed	2
Pantodonta		*Pantolambda*	Semi-lophed	4
		Titanoides	Semi-lophed	4
'Condylarthra'	Hyopsodontidae	*Haplaletes*	Bunodont	1
		Litomylus	Bunodont	1
	Mioclaenidae	*Ellipsodon*	Bunodont	1
		Litaletes	Bunodont	1
		Mioclaenus	Bunodont	2
		Promioclaenus	Bunodont	1
		Protoselene	Bunodont	1
	Periptychidae	*Anisonchus*	Bunodont	1
		Haploconus	Bunodont	1
		Periptychus	Bunodont	2
	Phenacodontidae	*Ectocion*	Semi-lophed	2
		Phenacodus	Bunodont	2
		Tetraclaenodon	Bunodont	2
Tiffanian 1				
Taeniodonta		*Conoryctes*	Bunodont	2
		Psittacotherium	Semi-lophed	2
Pantodonta		*Caenolambda*	Semi-lophed	4
		Pantolambda	Semi-lophed	4
		Titanoides	Semi-lophed	4
'Condylarthra'	Hyopsodontidae	*Haplaletes*	Bunodont	1
		Litomylus	Bunodont	1
	Mioclaenidae	*Litaletes*	Bunodont	1
		Promioclaenus	Bunodont	1
		Protoselene	Bunodont	1
	Periptychidae	*Anisonchus*	Bunodont	1
		Periptychus	Bunodont	2
	Phenacodontidae	*Ectocion*	Semi-lophed	2

PALEOCENE (*cont.*)

Tiffanian 1 (cont.)

		Phenacodus	Bunodont	2
		Tetraclaenodon	Bunodont	2

Tiffanian 2

Taeniodonta		*Psittacotherium*	Semi-lophed	2
Pantodonta		*Caenolambda*	Semi-lophed	4
		Titanoides	Semi-lophed	4
'Condylarthra'	Hyopsodontidae	*Haplaletes*	Bunodont	1
		Litomylus	Bunodont	1
	Mioclaenidae	*Promioclaenus*	Bunodont	1
		Protoselene	Bunodont	1
	Periptychidae	*Periptychus*	Bunodont	2
	Phenacodontidae	*Ectocion*	Semi-lophed	2
		Phenacodus	Bunodont	2

Tiffanian 3

Taeniodonta		*Psittacotherium*	Semi-lophed	2
Pantodonta		*Barylambda*	Semi-lophed	5
		Caenolambda	Semi-lophed	4
		Haplolambda	Semi-lophed	4
		Titanoides	Semi-lophed	4
Dinocerata		*Prodinoceras*	Semi-lophed	4
'Condylarthra'	Hyopsodontidae	*Aletodon*	Bunodont	1
		Dorraletes	Bunodont	1
		Haplaletes	Bunodont	1
		Litomylus	Bunodont	1
	Mioclaenidae	*Promioclaenus*	Bunodont	1
		Protoselene	Bunodont	1
	Periptychidae	*Periptychus*	Bunodont	2
	Phenacodontidae	*Ectocion*	Semi-lophed	2
		Phenacodus	Bunodont	2

Tiffanian 4

Taeniodonta		*Psittacotherium*	Semi-lophed	2
Pantodonta		*Barylambda*	Semi-lophed	5
		Caenolambda	Semi-lophed	4
		Haplolambda	Semi-lophed	4
		Ignatiolambda	Semi-lophed	4
		Titanoides	Semi-lophed	4
Dinocerata		*Prodinoceras*	Semi-lophed	4

PALEOCENE (*cont.*)

Tiffanian 4 (*cont.*)

'Condylarthra'	Hyopsodontidae	*Aletodon*	Bunodont	1
		Dorraletes	Bunodont	1
		Haplaletes	Bunodont	1
		Litomylus	Bunodont	1
		Utemylus	Bunodont	1
	Mioclaenidae	*Phenacodaptes*	Bunodont	1
	Periptychidae	*Periptychus*	Bunodont	2
	Phenacodontidae	*Ectocion*	Semi-lophed	2
		Phenacodus	Bunodont	2

Tiffanian 5

Taeniodonta		*Psittacotherium*	Semi-lophed	2
		Ectoganus	Semi-lophed	3
Pantodonta		*Barylambda*	Semi-lophed	5
		Caenolambda	Semi-lophed	4
		Haplolambda	Semi-lophed	4
		Ignatiolambda	Semi-lophed	4
		Leptolambda	Semi-lophed	5
		Titanoides	Semi-lophed	4
Dinocerata		*Prodinoceras*	Semi-lophed	4
'Condylarthra'	Hyopsodontidae	*Aletodon*	Bunodont	1
		Dorraletes	Bunodont	1
		Haplaletes	Bunodont	1
		Litomylus	Bunodont	1
	Mioclaenidae	*Phenacodaptes*	Bunodont	1
	Phenacodontidae	*Ectocion*	Semi-lophed	2
		Phenacodus	Bunodont	2

Tiffanian 6

Taeniodonta		*Psittacotherium*	Semi-lophed	2
		Ectoganus	Semi-lophed	3
Tillodonta		*Esthonyx*	Semi-lophed	2
Pantodonta		*Barylambda*	Semi-lophed	5
		Caenolambda	Semi-lophed	4
		Haplolambda	Semi-lophed	4
		Leptolambda	Semi-lophed	5
		Titanoides	Semi-lophed	4
Dinocerata		*Prodinoceras*	Semi-lophed	4
'Condylarthra'	Hyopsodontidae	*Aletodon*	Bunodont	1
	Mioclaenidae	*Apheliscus*	Bunodont	1
		Phenacodaptes	Bunodont	1

PALEOCENE (*cont.*)

Tiffanian 6 (*cont.*)

	Phenacodontidae	*Ectocion*	Semi-lophed	2
		Phenacodus	Bunodont	2
Arctostylopoidea		*Arctostylops*	Lophed	1

Clarkforkian 1

Taeniodonta		*Ectoganus*	Semi-lophed	3
Tillodonta		*Esthonyx*	Semi-lophed	2
Pantodonta	Barylambdidae	*Barylambda*	Semi-lophed	5
		Caenolambda	Semi-lophed	4
		Haplolambda	Semi-lophed	4
		Leptolambda	Semi-lophed	5
		Titanoides	Semi-lophed	4
	Coryphodontidae	*Coryphodon*	Semi-lophed	4
Dinocerata		*Prodinoceras*	Semi-lophed	4
'Condylarthra'	Hyopsodontidae	*Aletodon*	Bunodont	1
		Haplomylus	Bunodont	1
	Mioclaenidae	*Apheliscus*	Bunodont	1
		Phenacodaptes	Bunodont	1
	Phenacodontidae	*Ectocion*	Semi-lophed	2
		Phenacodus	Bunodont	2
Arctostylopoidea		*Arctostylops*	Lophed	1

Clarkforkian 2

Taeniodonta		*Ectoganus*	Semi-lophed	3
Tillodonta		*Esthonyx*	Semi-lophed	2
Pantodonta	Barylambdidae	*Caenolambda*	Semi-lophed	4
		Haplolambda	Semi-lophed	4
	Coryphodontidae	*Coryphodon*	Semi-lophed	4
Dinocerata		*Prodinoceras*	Semi-lophed	4
'Condylarthra'	Hyopsodontidae	*Aletodon*	Bunodont	1
		Haplomylus	Bunodont	1
	Mioclaenidae	*Apheliscus*	Bunodont	1
		Phenacodaptes	Bunodont	1
	Phenacodontidae	*Copecion*	Bunodont	2
		Ectocion	Semi-lophed	2
		Phenacodus	Bunodont	2
Arctostylopoidea		*Arctostylops*	Lophed	1

Clarkforkian 3

Taeniodonta		*Ectoganus*	Semi-lophed	3
Tillodonta		*Esthonyx*	Semi-lophed	2

PALEOCENE (*cont.*)

Clarkforkian 3 (cont.)

Pantodonta	Coryphodontidae	*Coryphodon*	Semi-lophed	4
Dinocerata		*Prodinoceras*	Semi-lophed	4
'Condylarthra'	Hyopsodontidae	*Aletodon*	Bunodont	1
		Haplomylus	Bunodont	1
		Hyopsodus	Bunodont	1
	Mioclaenidae	*Apheliscus*	Bunodont	1
	Phenacodontidae	*Copecion*	Bunodont	2
		Ectocion	Semi-lophed	2
		Meniscotherium	Lophed	2
		Phenacodus	Bunodont	2
Arctostylopoidea		*Arctostylops*	Lophed	1

EOCENE

Early Wasatchian

Taeniodonta		*Ectoganus*	Semi-lophed	3
Tillodonta		*Esthonyx*	Semi-lophed	2
Pantodonta	Coryphodontidae	*Coryphodon*	Semi-lophed	4
Dinocerata		*Prodinoceras*	Semi-lophed	4
'Condylarthra'	Hyopsodontidae	*Haplomylus*	Bunodont	1
		Hyopsodus	Bunodont	1
	Mioclaenidae	*Apheliscus*	Bunodont	1
	Phenacodontidae	*Copecion*	Bunodont	2
		Ectocion	Semi-lophed	2
		Meniscotherium	Lophed	2
		Phenacodus	Bunodont	2
Perissodactyla	Equidae	*Hyracotherium*	Semi-lophed	2
		Xenicohippus	Semi-lophed	2
	'Tapiroidea'	*Cardiolophus*	Semi-lophed	2
		Heptodon	Lophed	2
		Homogalax	Semi-lophed	2
Artiodactyla	'Dichobunidae'	*Bunophorus*	Bunodont	2
		Diacodexis	Bunodont	1
		'*Homacodon*' n.g.	Bunodont	1
		Simpsonodus	Bunodont	1

Middle Wasatchian

Taeniodonta		*Ectoganus*	Semi-lophed	3
Tillodonta		*Esthonyx*	Semi-lophed	2

EOCENE (*cont.*)

Middle Wasatchian (*cont.*)				
Pantodonta	Coryphodontidae	*Coryphodon*	Semi-lophed	4
Dinocerata		*Prodinoceras*	Semi-lophed	4
'Condylarthra'	Hyopsodontidae	*Haplomylus*	Bunodont	1
		Hyopsodus	Bunodont	1
	Mioclaenidae	*Apheliscus*	Bunodont	1
	Phenacodontidae	*Copecion*	Bunodont	2
		Ectocion	Semi-lophed	2
		Meniscotherium	Lophed	2
		Phenacodus	Bunodont	2
Perissodactyla	Equidae	*Hyracotherium*	Semi-lophed	2
		Xenicohippus	Semi-lophed	2
	'Tapiroidea'	*Cardiolophus*	Semi-lophed	2
		Heptodon	Lophed	2
		Homogalax	Semi-lophed	2
		Hyrachyus	Lophed	3
		Paleomoropus	Semi-lophed	2
Artiodactyla	'Dichobunidae'	*Bunophorus*	Bunodont	2
		Diacodexis	Bunodont	1
		Hexacodus	Semi-lophed	1
		Simpsonodus	Bunodont	1
Late Wasatchian				
Taeniodonta		*Ectoganus*	Semi-lophed	3
		Stylinodon	Semi-lophed	3
Tillodonta		*Esthonyx*	Semi-lophed	2
		Megalesthonyx	Semi-lophed	3
		Trogosus	Semi-lophed	4
Pantodonta	Coryphodontidae	*Coryphodon*	Semi-lophed	4
Dinocerata		*Bathyopsis*	Semi-lophed	4
		Prodinoceras	Semi-lophed	4
'Condylarthra'	Hyopsodontidae	*Haplomylus*	Bunodont	1
		Hyopsodus	Bunodont	1
	Mioclaenidae	*Apheliscus*	Bunodont	1
	Phenacodontidae	*Copecion*	Bunodont	2
		Ectocion	Semi-lophed	2
		Meniscotherium	Lophed	2
		Phenacodus	Bunodont	2
Perissodactyla	Equidae	*Hyracotherium*	Semi-lophed	2
		Orohippus	Semi-lophed	2
	'Tapiroidea'	*Cardiolophus*	Semi-lophed	2

EOCENE (*cont.*)

Late Wasatchian (*cont.*)

		Heptodon	Lophed	2
		Homogalax	Semi-lophed	2
		Hyrachyus	Lophed	3
		Selenaletes	Semi-lophed	2
	Brontotheriidae	*Eotitanops*	Semi-lophed	3
	incertae sedis	*Lambdotherium*	Semi-lophed	3
Artiodactyla	'Dichobunidae'	*Antiacodon*	Semi-lophed	1
		Bunophorus	Bunodont	2
		Diacodexis	Bunodont	1
		'D.' waltonensis	Bunodont	1
		Hexacodus	Semi-lophed	1

Early Bridgerian

Taeniodonta		*Stylinodon*	Semi-lophed	3
Tillodonta		*Esthonyx*	Semi-lophed	2
		Trogosus	Semi-lophed	4
Pantodonta	Coryphodontidae	*Coryphodon*	Semi-lophed	4
Dinocerata		*Bathyopsis*	Semi-lophed	4
'Condylarthra'	Hyopsodontidae	*Hyopsodus*	Bunodont	1
	Phenacodontidae	*Ectocion*	Semi-lophed	2
		Meniscotherium	Lophed	2
		Phenacodus	Bunodont	2
Perissodactyla	Equidae	*Hyracotherium*	Semi-lophed	2
		Orohippus	Semi-lophed	2
	'Tapiroidea'	*Fouchia*	Lophed	2
		Helaletes	Lophed	2
		Heptodon	Lophed	2
		Hyrachyus	Lophed	3
		Isectolophus	Lophed	2
		Selenaletes	Semi-lophed	2
	Brontotheriidae	*Eotitanops*	Semi-lophed	3
		Palaeosyops	Semi-lophed	4
Artiodactyla	'Dichobunidae'	*Antiacodon*	Semi-lophed	1
		Bunophorus	Bunodont	2
		Diacodexis	Bunodont	1
		'D.' waltonensis	Bunodont	1
		Helohyus	Bunodont	2
		Homacodon	Semi-lophed	1
		Microsus	Bunodont	1

EOCENE (*cont.*)				
Middle Bridgerian				
Taeniodonta		*Stylinodon*	Semi-lophed	3
Tillodonta		*Tillodon*	Semi-lophed	4
		Trogosus	Semi-lophed	4
Dinocerata		*Bathyopsis*	Semi-lophed	4
		Uintatherium	Semi-lophed	5
'Condylarthra'	Hyopsodontidae	*Hyopsodus*	Bunodont	1
	Phenacodontidae	*Meniscotherium*	Lophed	2
		Phenacodus	Bunodont	2
Perissodactyla	Equidae	*Orohippus*	Semi-lophed	2
	'Tapiroidea'	*Dilophodon*	Lophed	2
		Helaletes	Lophed	2
		Hyrachyus	Lophed	3
		Isectolophus	Lophed	2
	Brontotheriidae	*Palaeosyops*	Semi-lophed	4
Artiodactyla	'Dichobunidae'	*Achaenodon*	Bunodont	3
		Antiacodon	Semi-lophed	1
		'D.' waltonensis	Bunodont	1
		Helohyus	Bunodont	2
		Homacodon	Semi-lophed	1
		Microsus	Bunodont	1
Late Bridgerian				
Taeniodonta		*Stylinodon*	Semi-lophed	3
Dinocerata		*Tethyopsis*	Semi-lophed	5
		Uintatherium	Semi-lophed	5
'Condylarthra'	Hyopsodontidae	*Hyopsodus*	Bunodont	1
	Phenacodontidae	*Phenacodus*	Bunodont	2
Perissodactyla	Equidae	*Epihippus*	Semi-lophed	2
		Orohippus	Semi-lophed	2
	'Tapiroidea'	*Desmatotherium*	Lophed	4
		Dilophodon	Lophed	2
		Hyrachyus	Lophed	3
		Isectolophus	Lophed	2
	Rhinoceratoidea	*Amynodon*	Lophed	5
	Brontotheriidae	*Palaeosyops*	Semi-lophed	4
		Mesatirhinus	Semi-lophed	5
		Telmatherium	Semi-lophed	5
Artiodactyla	'Dichobunidae'	*Achaenodon*	Bunodont	3
		Antiacodon	Semi-lophed	1

EOCENE (*cont.*)

Late Bridgerian (*cont.*)

		Helohyus	Bunodont	2
		Homacodon	Semi-lophed	1
		Microsus	Bunodont	1
		Neodiacodexis	Bunodont	1
		Sarcolemur	Bunodont	1
	Agriochoeridae	*Protoreodon*	Semi-lophed	2

Early Uintan

Taeniodonta		*Stylinodon*	Semi-lophed	3
Dinocerata		*Eobasilus*	Semi-lophed	5
		Tethyopsis	Semi-lophed	5
		Uintatherium	Semi-lophed	5
'Condylarthra'	Hyopsodontidae	*Hyopsodus*	Bunodont	1
Perissodactyla	Equidae	*Epihippus*	Semi-lophed	2
		Orohippus	Semi-lophed	2
	'Tapiroidea'	*Colodon*	Lophed	3
		Dilophodon	Lophed	2
		New genus	Lophed	2
		Helaletes	Lophed	2
		Hyrachyus	Lophed	3
		Isectolophus	Lophed	2
	Rhinoceratoidea	*Amynodon*	Lophed	5
		Epitriplopus	Lophed	2
		Triplopus	Lophed	2
		Uintaceras	Lophed	4
	Brontotheriidae	*Metarhinus*	Semi-lophed	5
		Sphenocoelus	Semi-lophed	5
		Sthenodectes	Semi-lophed	5
	Chalicotherioidea	*Eomoropus*	Lophed	2
		Grangeria	Lophed	3
Artiodactyla	'Dichobunidae'	*Achaenodon*	Bunodont	3
		Bunomeryx	Semi-lophed	1
		Helohyus	Bunodont	2
		'*Helohyus*' n.g.	Bunodont	2
		Hylomeryx	Semi-lophed	1
		'*Homacodon*' n.g.	Semi-lophed	1
		Laredochoerus	Bunodont	2
		Parahyus	Bunodont	3
	Suoidea *inc.sed.*	*Brachyhyops*	Bunodont	2
	Agriochoeridae	*Agriochoerus*	Lophed	3

EOCENE (*cont.*)
Early Uintan (*cont.*)

		Protoreodon	Semi-lophed	2
	Protoceratidae	*Leptoreodon*	Semi-lophed	2
		Leptotragulus	Semi-lophed	2
	Oromerycidae	*Oromeryx*	Semi-lophed	2
		Malaquiferus	Semi-lophed	2
		Merycobunodon	Semi-lophed	2
		Protylopus	Semi-lophed	2
	Camelidae	*Poebrodon*	Lophed	2
Late Uintan				
Taeniodonta		*Stylinodon*	Semi-lophed	3
Dinocerata		*Uintatherium*	Semi-lophed	5
'Condylarthra'	Hyopsodontidae	*Hyopsodus*	Bunodont	1
Perissodactyla	Equidae	*Epihippus*	Semi-lophed	2
	'Tapiroidea'	*Colodon*	Lophed	3
		Helaletes	Lophed	2
		Heteraletes	Lophed	3
		Hyrachyus	Lophed	3
		Isectolophus	Lophed	2
		Protapirus	Lophed	4
		Schizotheriodes	Lophed	2
		New genus	Lophed	2
	Rhinoceratoidea	*Amynodon*	Lophed	5
		Amynodontopsis	Lophed	5
		Epitriplopus	Lophed	2
		Megalamynodon	Lophed	5
		Metamynodon	Lophed	5
		Triplopus	Lophed	2
		Uintaceras	Lophed	4
	Brontotheriidae	*'Diplacodon'*	Semi-lophed	5
		Eotitanotherium	Semi-lophed	5
		Metatelmatherium	Semi-lophed	5
		Notiotitanops	Semi-lophed	5
		Protitanotherium	Semi-lophed	5
	Chalicotherioidea	*Grangeria*	Lophed	3
Artiodactyla	'Dichobunidae'	*Apriculus*	Semi-lophed	2
		Auxontodon	Bunodont	2
		Bunomeryx	Semi-lophed	1
		Hylomeryx	Semi-lophed	1
		Ibarus	Bunodont	1

EOCENE (*cont.*)
Late Uintan (*cont.*)

		Mesomeryx	Semi-lophed	1
		Mytonomeryx	Semi-lophed	2
		Pentacemylus	Semi-lophed	2
		Tapochoerus	Bunodont	2
		Texodus	Semi-lophed	1
	Suoidea *inc.sed.*	*Brachyhyops*	Bunodont	2
	Agriochoeridae	*Agriochoerus*	Lophed	3
		Protoreodon	Semi-lophed	2
		'P.' n.g. A	Semi-lophed	1
		'P.' n.g. B	Semi-lophed	1
		'P.' n.g. C	Semi-lophed	1
	Protoceratidae	*Heteromeryx*	Semi-lophed	2
		Leptoreodon	Semi-lophed	2
		Leptotragulus	Semi-lophed	2
		Poabromylus	Semi-lophed	2
		Toromeryx	Semi-lophed	2
	Oromerycidae	*Oromeryx*	Semi-lophed	2
		Malaquiferus	Semi-lophed	2
		Protylopus	Semi-lophed	2
	Camelidae	*Poebrodon*	Lophed	2
	Hypertragulidae	*Simimeryx*	Semi-lophed	1
	Leptomerycidae	*Hendryomeryx*	Semi-lophed	1
		Leptomeryx	Semi-lophed	2
Duchesnean				
'Condylarthra'	Hyopsodontidae	*Hyopsodus*	Bunodont	1
Perissodactyla	Equidae	*Epihippus*	Semi-lophed	2
		Haplohippus	Semi-lophed	2
		Mesohippus	Lophed	3
	'Tapiroidea'	*Colodon*	Lophed	3
		Heteraletes	Lophed	3
		Protapirus	Lophed	4
		Toxotherium	Lophed	3
	Rhinoceratoidea	*Amynodontopsis*	Lophed	5
		Epitriplopus	Lophed	2
		Hyracodon	Lophed	3
		Megalamynodon	Lophed	5
		Metamynodon	Lophed	5
		Penetrigonias	Lophed	4
		'Procardurcodon'	Lophed	5

EOCENE (*cont.*)
Duchesnean (*cont.*)

		Teleteceras	Lophed	4
		Triplopus	Lophed	2
		Uintaceras	Lophed	4
	Brontotheriidae	*Duchesneodus*	Semi-lophed	5
		Protitanops	Semi-lophed	5
Artiodactyla	'Dichobunidae'	*Discritocheorus*	Bunodont	2
		Leptochoerid n.g.	Bunodont	2
		Pentacemylus	Semi-lophed	2
	Anthracotheriidae	*Heptacodon*	Semi-lophed	3
	Suoidea *inc.sed.*	*Brachyhyops*	Bunodont	2
	Agriochoeridae	*Agriochoerus*	Lophed	3
		'A.' n.g. E	Lophed	3
		'A.' n.g. F	Lophed	3
		Protoreodon	Semi-lophed	2
		'P.' n.g. A	Semi-lophed	1
		'P.' n.g. B	Semi-lophed	1
		'P.' n.g. C	Semi-lophed	1
		'P' n.g. D	Semi-lophed	2
	Merycoidodontidae	*Aclistomycter*	Lophed	2
		Bathygenus	Lophed	2
		Leptauchenia	Lophed	3
		Oreonetes	Lophed	2
		Prodesmatochoerus	Lophed	3
	Protoceratidae	*Heteromeryx*	Semi-lophed	2
		Leptoreodon	Semi-lophed	2
		Leptotragulus	Semi-lophed	2
		Poabromylus	Semi-lophed	2
		Pseudoprotoceras	Lophed	2
	Oromerycidae	*Eotylopus*	Semi-lophed	2
		Oromeryx	Semi-lophed	2
		Protylopus	Semi-lophed	2
	Camelidae	*Poebrodon*	Lophed	2
	Hypertragulidae	*Hypertragulus*	Semi-lophed	2
		Hypisodus	Lophed	1
		Parvitragulus	Semi-lophed	1
		Simimeryx	Semi-lophed	1
	Leptomerycidae	*Hendryomeryx*	Semi-lophed	1
		Leptomeryx	Semi-lophed	2
	Ruminantia *inc.sed.*	*Hidrosotherium*	Semi-lophed	2

EOCENE (*cont.*)

Early Chadronian

'Condylarthra'	Hyopsodontidae	*Hyopsodus*	Bunodont	1
Perissodactyla	Equidae	*Mesohippus*	Lophed	3
		Miohippus	Lophed	3
	'Tapiroidea'	*Colodon*	Lophed	3
		Protapirus	Lophed	4
		Toxotherium	Lophed	3
	Rhinoceratoidea	*Amphicaenopus*	Lophed	5
		Hyracodon	Lophed	3
		Metamynodon	Lophed	5
		Penetrigonias	Lophed	4
		Subhyracodon	Lophed	4
		Triplopides	Lophed	4
		Trigonias	Lophed	5
	Brontotheriidae	*Brontops*	Semi-lophed	5
		Megacerops	Semi-lophed	5
		Menops	Semi-lophed	5
Artiodactyla	'Dichobunidae'	*Stibarus*	Bunodont	1
	Entelodontidae	*Archaeotherium*	Bunodont	4
	Anthracotheriidae	*Aepinacodon*	Semi-lophed	4
		Bothriodon	Semi-lophed	4
		Heptacodon	Semi-lophed	3
	Suoidea *inc.sed.*	*Brachyhyops*	Bunodont	2
	Agriochoeridae	*Agriochoerus*	Lophed	3
		'A.' n.g. E	Lophed	3
		Protoreodon	Semi-lophed	2
		'P.' n.g. B	Semi-lophed	1
		'P.' n.g. C	Semi-lophed	1
	Merycoidodontidae	*Bathygenus*	Lophed	2
		Leptauchenia	Lophed	3
		Oreonetes	Lophed	2
		Prodesmatochoerus	Lophed	3
	Protoceratidae	*Heteromeryx*	Semi-lophed	2
		'Leptotragulus'	Semi-lophed	2
		Poabromylus	Semi-lophed	2
		Pseudoprotoceras	Lophed	2
	Oromerycidae	*Eotylopus*	Semi-lophed	2
		Montanatylopus	Semi-lophed	3
	Camelidae	*Poebrotherium*	Lophed	2
		'P.' *franki*	Lophed	2
	Hypertragulidae	*Hypertragulus*	Semi-lophed	2

EOCENE (*cont.*)

Early Chadronian (*cont.*)

		Hypisodus	Lophed	1
		Parvitragulus	Semi-lophed	1
	Leptomerycidae	*Hendryomeryx*	Semi-lophed	1
		Leptomeryx	Semi-lophed	2

Middle Chadronian

Perissodactyla	Equidae	*Mesohippus*	Lophed	3
		Miohippus	Lophed	3
	'Tapiroidea'	*Colodon*	Lophed	3
		Protapirus	Lophed	4
		Toxotherium	Lophed	3
	Rhinoceratoidea	*Amphicaenopus*	Lophed	5
		Hyracodon	Lophed	3
		Metamynodon	Lophed	5
		Penetrigonias	Lophed	4
		Subhyracodon	Lophed	4
		Trigonias	Lophed	5
	Brontotheriidae	*Brontops*	Semi-lophed	5
		Megacerops	Semi-lophed	5
		Menops	Semi-lophed	5
Artiodactyla	'Dichobunidae'	*Stibarus*	Bunodont	1
	Entelodontidae	*Archaeotherium*	Bunodont	4
	Anthracotheriidae	*Aepinacodon*	Semi-lophed	4
		Bothriodon	Semi-lophed	4
		Heptacodon	Semi-lophed	3
	Agriochoeridae	*Agriochoerus*	Lophed	3
	Merycoidodontidae	*Bathygenus*	Lophed	2
		Leptauchenia	Lophed	3
		Oreonetes	Lophed	2
		Prodesmatochoerus	Lophed	3
	Protoceratidae	*Heteromeryx*	Semi-lophed	2
		'Leptotragulus'	Semi-lophed	2
		Poabromylus	Semi-lophed	2
		Pseudoprotoceras	Lophed	2
	Oromerycidae	*Eotylopus*	Semi-lophed	2
	Camelidae	*Poebrotherium*	Lophed	2
		'P.' franki	Lophed	2
	Hypertragulidae	*Hypertragulus*	Semi-lophed	2
		Hypisodus	Lophed	1
		Parvitragulus	Semi-lophed	1

EOCENE (*cont.*)

Middle Chadronian (*cont.*)

	Leptomerycidae	*Hendryomeryx*	Semi-lophed	1
		Leptomeryx	Semi-lophed	2

Late Chadronian

Perissodactyla	Equidae	*Mesohippus*	Lophed	3
		Miohippus	Lophed	3
	'Tapiroidea'	*Colodon*	Lophed	3
		Protapirus	Lophed	4
	Rhinoceratoidea	*Amphicaenopus*	Lophed	5
		Hyracodon	Lophed	3
		Metamynodon	Lophed	5
		Penetrigonias	Lophed	4
		Subhyracodon	Lophed	4
		Trigonias	Lophed	5
	Brontotheriidae	*Brontops*	Semi-lophed	5
		Megacerops	Semi-lophed	5
		Menops	Semi-lophed	5
Artiodactyla	'Dichobunidae'	*Stibarus*	Bunodont	1
		Leptochoerus	Bunodont	2
	Entelodontidae	*Archaeotherium*	Bunodont	4
	Anthracotheriidae	*Aepinacodon*	Semi-lophed	4
		Bothriodon	Semi-lophed	4
		Heptacodon	Semi-lophed	3
	Agriochoeridae	*Agriochoerus*	Lophed	3
	Merycoidodontidae	*Leptauchenia*	Lophed	3
		Oreonetes	Lophed	2
		Prodesmatochoerus	Lophed	3
	Protoceratidae	*Heteromeryx*	Semi-lophed	2
		'Leptotragulus'	Semi-lophed	2
		Poabromylus	Semi-lophed	2
		Pseudoprotoceras	Lophed	2
	Oromerycidae	*Eotylopus*	Semi-lophed	2
	Camelidae	*Poebrotherium*	Lophed	2
		Paratylopus	Lophed	2
	Hypertragulidae	*Hypertragulus*	Semi-lophed	2
		Hypisodus	Lophed	1
	Leptomerycidae	*Hendryomeryx*	Semi-lophed	1
		Leptomeryx	Semi-lophed	2

OLIGOCENE

Orellan				
Perissodactyla	Equidae	*Mesohippus*	Lophed	3
		Miohippus	Lophed	3
	'Tapiroidea'	*Colodon*	Lophed	3
		Protapirus	Lophed	4
	Rhinoceratoidea	*Amphicaenopus*	Lophed	5
		Hyracodon	Lophed	3
		Metamynodon	Lophed	5
		Penetrigonias	Lophed	4
		Subhyracodon	Lophed	4
		Trigonias	Lophed	5
Artiodactyla	'Dichobunidae'	*Stibarus*	Bunodont	1
		Leptochoerus	Bunodont	2
	Entelodontidae	*Archaeotherium*	Bunodont	4
	Anthracotheriidae	*Bothriodon*	Semi-lophed	4
		Elomeryx	Semi-lophed	4
		Heptacodon	Semi-lophed	3
	Tayassuidae	*Perchoerus*	Bunodont	2
		Thinohyus	Bunodont	2
	Agriochoeridae	*Agriochoerus*	Lophed	3
	Merycoidodontidae	*Eucrotaphus*	Lophed	3
		Leptauchenia	Lophed	3
		Oreonetes	Lophed	2
		Prodesmatochoerus	Lophed	3
	Protoceratidae	*Pseudoprotoceras*	Lophed	2
	Camelidae	*Poebrotherium*	Lophed	2
		Paratylopus	Lophed	2
	Hypertragulidae	*Hypertragulus*	Semi-lophed	2
		Hypisodus	Lophed	1
	Leptomerycidae	*Hendryomeryx*	Semi-lophed	1
		Leptomeryx	Semi-lophed	2
Whitneyan				
Perissodactyla	Equidae	*Mesohippus*	Lophed	3
		Miohippus	Lophed	3
	'Tapiroidea'	*Colodon*	Lophed	3
		Protapirus	Lophed	4
	Rhinoceratoidea	*Amphicaenopus*	Lophed	5
		Hyracodon	Lophed	3
		Metamynodon	Lophed	5

OLIGOCENE (*cont.*)
Whitneyan (*cont.*)

		Penetrigonias	Lophed	4
		Subhyracodon	Lophed	4
		Diceratherium	Lophed	5
Artiodactyla	'Dichobunidae'	*Stibarus*	Bunodont	1
		Leptochoerus	Bunodont	2
	Entelodontidae	*Archaeotherium*	Bunodont	4
		Choerodon	Bunodont	5
		Megachoerus	Bunodont	5
	Anthracotheriidae	*Elomeryx*	Semi-lophed	4
		Heptacodon	Semi-lophed	3
	Tayassuidae	*Perchoerus*	Bunodont	2
		Thinohyus	Bunodont	2
	Agriochoeridae	*Agriochoerus*	Lophed	3
	Merycoidodontidae	*Blickohyus*	Lophed	3
		Eporeodon	Lophed	3
		Eucrotaphus	Lophed	3
		Leptauchenia	Lophed	3
		Oreonetes	Lophed	2
		Prodesmatochoerus	Lophed	3
	Protoceratidae	*Protoceras*	Lophed	3
	Camelidae	*Poebrotherium*	Lophed	2
		Paralabis	Lophed	3
		Paratylopus	Lophed	2
		Pseudolabis	Lophed	3
	Hypertragulidae	*Hypertragulus*	Semi-lophed	2
		Hypisodus	Lophed	1
		Nanotragulus	Lophed	2
	Leptomerycidae	*Hendryomeryx*	Semi-lophed	1
		Leptomeryx	Semi-lophed	2

Early Early Arikareean

Perissodactyla	Equidae	*Kalobatippus*	Lophed	3
		Mesohippus	Lophed	3
		Miohippus	Lophed	3
	Tapiridae	*Miotapirus*	Lophed	3
		Protapirus	Lophed	4
	Rhinoceratoidea	*Diceratherium*	Lophed	5
		Hyracodon	Lophed	3
		Subhyracodon	Lophed	4
Artiodactyla	'Dichobunidae'	*Leptochoerus*	Bunodont	2

OLIGOCENE (*cont.*)

Early Early Arikareean (cont.)

	Entelodontidae	*Archaeotherium*	Bunodont	2
		Choerodon	Bunodont	5
		Dinohyus	Bunodont	5
	Anthracotheriidae	*Elomeryx*	Semi-lophed	4
		Kukusepasutanka	Semi-lophed	4
	Tayassuidae	*Thinohyus*	Bunodont	2
	Agriochoeridae	*Agriochoerus*	Lophed	3
	Merycoidodontidae	*Blickohyus*	Lophed	3
		Eporeodon	Lophed	3
		Leptauchenia	Lophed	3
		Merycochoerus	Lophed	4
		Merycoides	Lophed	4
		Oreodontoides	Lophed	3
		Prodesmatochoerus	Lophed	3
		Sespia	Lophed	2
	Protoceratidae	*Protoceras*	Lophed	3
	Camelidae	*Gentilicamelus*	Lophed	2
		Miotylopus	Lophed	3
		Poebrotherium	Lophed	2
		Pseudolabis	Lophed	3
		Stenomylus	Lophed	3
	Hypertragulidae	*Hypertragulus*	Semi-lophed	2
		Hypisodus	Lophed	1
		Nanotragulus	Lophed	2
	Leptomerycidae	*Leptomeryx*	Semi-lophed	2
		Pronodens	Lophed	2

Late Early Arikareean

Perissodactyla	Equidae	*Kalobatippus*	Lophed	3
		Miohippus	Lophed	3
	Tapiridae	*Miotapirus*	Lophed	3
		Protapirus	Lophed	4
	Rhinoceratoidea	*Diceratherium*	Lophed	5
Artiodactyla	'Dichobunidae'	*Leptochoerus*	Bunodont	2
	Entelodontidae	*Archaeotherium*	Bunodont	2
		Dinohyus	Bunodont	5
	Anthracotheriidae	*Arretotherium*	Semi-lophed	4
	Tayassuidae	*Thinohyus*	Bunodont	2
	Agriochoeridae	*Agriochoerus*	Lophed	3
	Merycoidodontidae	*Eporeodon*	Lophed	3

OLIGOCENE (*cont.*)

Late Early Arikareean (*cont.*)

	Leptauchenia	Lophed	3
	Merycochoerus	Lophed	4
	Merycoides	Lophed	4
	Oreodontoides	Lophed	3
	Sespia	Lophed	2
Protoceratidae	*Protoceras*	Lophed	3
Camelidae	*Miotylopus*	Lophed	3
	Pseudolabis	Lophed	3
	Stenomylus	Lophed	3
Hypertragulidae	*Hypertragulus*	Semi-lophed	2
	Nanotragulus	Lophed	2
Leptomerycidae	*Leptomeryx*	Semi-lophed	2
	Pronodens	Lophed	2

BRUCE J. MACFADDEN

8

Origin and evolution of the grazing guild in Cenozoic New World terrestrial mammals

Introduction

Today grasslands cover about 25% of the world's land surface and constitute an enormous food resource that is exploited by invertebrate and vertebrate grazers. Grazers are defined as herbivores with diets consisting predominantly (>90%) of grass and other associated low ground cover in grassland biomes (Janis and Ehrhardt 1988). In present-day ecosystems, grazing is a common-place feeding strategy. Although terrestrial herbivory can be documented in vertebrates over the past 300 million years since the Carboniferous (Sues and Reisz 1998; Reisz and Sues, this volume), the origin of the grazing guild in vertebrates is a relatively recent arrival on the global ecological landscape.

This chapter presents paleontological evidence that terrestrial grazing guilds have only existed since the middle Tertiary, about 35 million years ago. I will confine my discussion here to mammals because: (1) although in terms of biomass and diversity, invertebrate grazers (mostly insects) are potentially the largest component of terrestrial grazing guilds worldwide (Dyer *et al.* 1982), the fossil record of this group is relatively poor; and (2) extinct mammalian grazers generally have an exceedingly rich fossil record that can be used to understand the evolution of the grazing guild. Furthermore, recent studies of extinct mammalian grazers combine a diverse array of traditional morphological evidence along with some new techniques, including stable isotopic analyses. Together these techniques allow a better understanding of the origin and evolution of the grazing guild. This chapter will focus attention on the fossil record in the New World; however, the pattern described here also is generally applicable to the Old World.

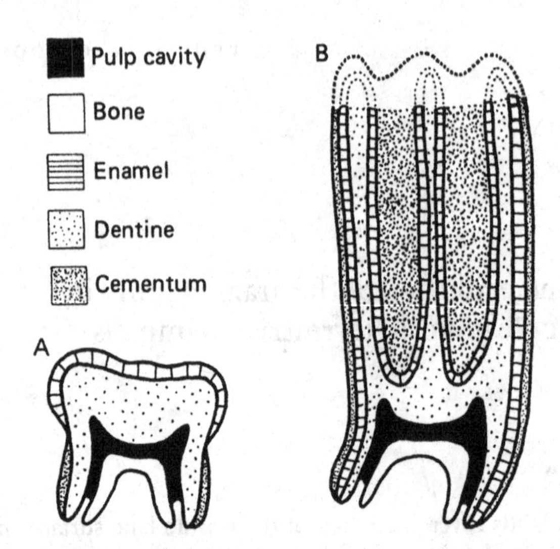

Figure 8.1. Cross-sections of (a) low-crowned or brachydont (human), and (b) high-crowned or hypsodont (horse) molar teeth showing development of the various dental tissues and elongated crown in the horse. (From Janis and Fortelius 1988, reproduced with permission of Cambridge University Press.)

The fossil record of grazing mammals: morphological and isotopic evidence

Kovalevsky (1873) published a classic monograph on fossil horses, which provides the foundation for our understanding of the evolution of grazing in that group. He proposed that the acquisition of high-crowned teeth in horses during the Miocene was an evolutionary adaptation to feeding on grasses. The reason for the correlation of high-crowned teeth with grazing is because grasses are highly abrasive and tend to accelerate tooth wear relative to softer browse (Figure 8.1). Grasses contain microscopic, dumbbell-shaped particles called phytoliths (Figure 8.2) composed of silica (SiO_2). Although phytoliths may also function in support (e.g., in bamboo), it is fairly well accepted that the primary role of these particles is to deter herbivory. As such, the 'cost' of becoming a grazer is the evolutionary necessity to acquire progressively higher-crowned teeth. Thus, even though the teeth of grazers wear very rapidly, they have more tooth volume to wear down during their lifetime relative to those of a short-crowned browser feeding on softer plants (Figure 8.1).

There are two general evolutionary morphotypes related to acquiring high-crowned teeth for grazing. Hypsodont teeth are high-crowned

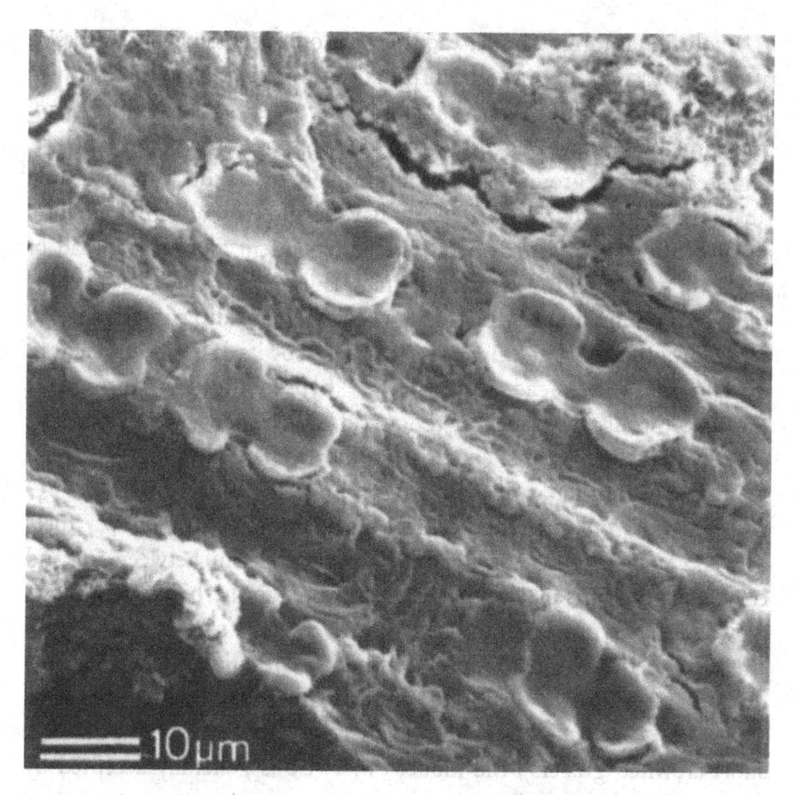

Figure 8.2. Dumbbell-shaped, abrasive silica particles (phytoliths) preserved in Miocene grass from Kenya. (From Dugas and Retallack 1993, reproduced with permission of The Paleontological Society.)

and have determinate growth. In addition to high crowns, hypsodont mammals further increase the durability of their teeth by infolding the enamel, which results in increased surface area for this more resistant kind of dental tissue (Van Valen 1960; Janis and Fortelius 1988). Many groups of extant or extinct presumed grazers are hypsodont, including from the New World the classic example of the horse, and also certain groups of rhinos, proboscideans, and ruminant artiodactyls. The other kind of high-crowned tooth is termed hypselodont, or ever-growing. Hypselodont teeth usually have relatively simple enamel patterns on the occlusal surface, and sometimes (e.g., in notoungulates), enamel is reduced in thickness or partially lost. Hypselodont teeth are mostly found in small North American mammals (rodents and lagomorphs) and are widespread among the extinct South American notoungulates.

Evolution is rarely simple, and there are other contributing factors or alternate reasons that have been proposed to explain the evolution of high-crowned teeth in Tertiary mammals. Workers such as White (1959) and Janis (1988) have suggested that the presence of contaminant grit ingested from the ground when cropping close to the substrate, or as fine dust on the grass fodder itself, could have resulted in, or contributed to, the evolution of high-crowned teeth independent from the spread of grasslands. Although it is possible that, as in the past (as also happens today), contaminant grit did get ingested during feeding on grasses and other low ground cover, this abrasive material by itself is not sufficient to explain the widespread acquisition of hypsodont teeth during the Miocene throughout much of the world. Consequently, most workers still favor as the principal explanation the acquisition of high-crowned teeth as a coevolutionary response to feeding primarily on grass.

Evidence from modern herbivorous mammals with known diets further confirms this classic hypothesis of a correlation between diet and crown height (Janis 1988). As a general rule, extant herbivores with low-crowned teeth are predominately browsers and species with high-crowned teeth are predominantly grazers. Although there are some notable exceptions to this oversimplified rule (e.g., the hippopotamus *Hippopotamus amphibius* and the llama *Lama glama* are both relatively short-crowned grazers), the model otherwise is generally accepted by most paleontologists.

Within the past decade, a new technique, stable isotopic analysis of tooth enamel, has been applied to the study of diets and ecology of extinct herbivorous mammals. CO_3^{2-} is incorporated into the hydroxyapatite crystal lattice of skeletal tissue (Newesley 1989; McClellan and Kauwenbergh 1990), primarily replacing the PO_4^{2-}, although lesser amounts also can weakly bond at other sites (OH^-). Carbon isotope analysis was initially applied to study diets of ancient human cultures using different kinds of skeletal tissues, including organic collagen and apatite, the latter of which is the mineral phase of bone and teeth (e.g., van der Merwe 1982; DeNiro 1987). Subsequent paleontological studies of stable isotopes analyzed mineralized bone, because collagen is rarely preserved in the fossil record. It was quickly found that the porous nature of bone makes it prone to post-mortem chemical alteration by ground water in the sediments. Thus, in the late 1980s, several workers began to concentrate their efforts on analyzing the stable isotopes of tooth enamel, which consists of more than 95% mineral hydroxyapatite (Hillson 1986) and is compact in its crystal structure and relatively resistant to diagenesis (e.g., Lee-Thorp and van der Merwe 1987; Quade *et al.* 1992). The analysis of the carbon isotopes in tooth enamel has opened the door to a major new line of paleontological investigation.

The stable isotopes of carbon (^{12}C and ^{13}C) are combined (fractionated) in different proportions depending upon whether a plant photosynthesizes this element using the Calvin, Hatch–Slack, or CAM pathways. Most (*c.* 85%) terrestrial plant biomass, including trees, shrubs, forbs, and cool-growing season grasses, photosynthesize carbon using the Calvin pathway. These are also called C_3 plants because of the formation of initial 3-carbon compounds during photosynthesis. Most tropical and temperate grasses photosynthesize carbon using the Hatch–Slack pathway; these are called C_4 plants because of the formation of 4-carbon compounds as the initial photosynthetic product. CAM, or Crassulacean Acid Metabolism, plants consist mostly of succulents such as the Cactacaea and have $\delta^{13}C$ values with a broad range that overlaps pure C_3 (infrequently) on the one hand, to pure C_4 (more frequently) on the other, depending upon the species. These will not be considered further in this chapter because they are not central to the origin of grasslands and the grazing guild.

C_3 and C_4 plants can be discriminated by their ratios of stable carbon isotopes, as expressed by the following standard notation:

$$\delta^{13}C_{PDB} \text{ (in parts per thousand, ‰)} = [(R_{sample}/R_{standard})-1] \times 1000$$

where $R = {}^{13}C/{}^{12}C$. All measurements of a fossil sample (in this case, tooth enamel) are compared with the international standard, the Cretaceous Pee Dee belemnite (PDB).

The two principal photosynthetic pathways of interest here fractionate carbon isotopes in fundamentally different proportions. C_3 plants have mean $\delta^{13}C$ values of −27‰ with a broad range between about −31 to −23‰ (Figure 8.3). In contrast, C_4 plants are isotopically more positive with mean $\delta^{13}C$ values of −13‰ with a more narrow range between about −15 and −11‰ (Deines 1980; Farquhar, Ehleringer and Hubick 1989; Boutton 1991). Relative to the $\delta^{13}C$ values of plant foodstuffs, tooth enamel carbonate of mammalian herbivores is enriched (i.e., less negative) by about 13‰ (Koch, Zachos and Gingerich 1992; Quade *et al.* 1992; Cerling *et al.* 1997).

Using the two methods (crown height and carbon isotopes) allows a better discrimination of the feeding ecologies of extinct herbivores. As shown in the lower left-hand quadrant of Figure 8.4, short-crowned

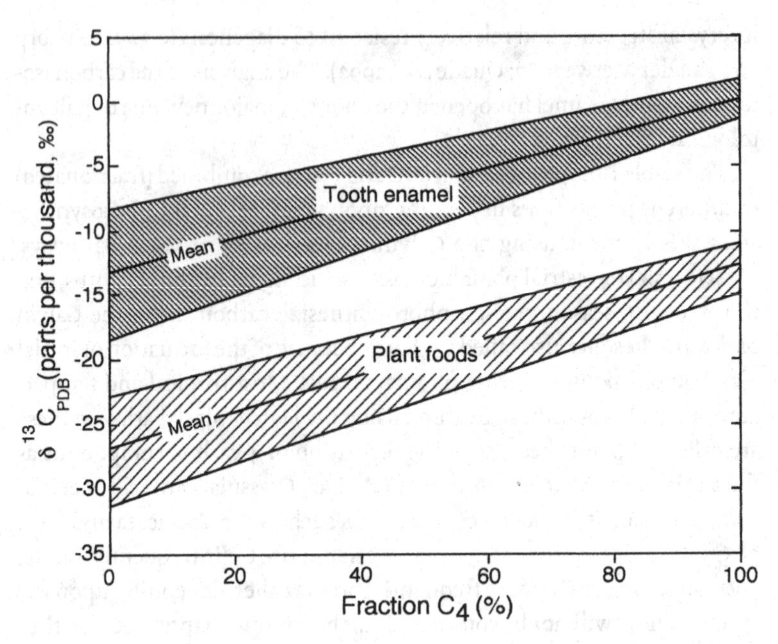

Figure 8.3. Plot of $\delta^{13}C$ values versus fraction C_4 contained in varied kinds of plants and the corresponding enriched (more positive) values of tooth enamel. This model assumes: (1) a mean $\delta^{13}C$ value for C_3 plants of −27‰ with a range of ± 4‰; (2) a mean $\delta^{13}C$ value for C_4 plants of −13‰ with a range of ± 2‰; and (3) relative to plant foodstuffs, fossil tooth enamel $\delta^{13}C$ (hydroxyapatite phase) is enriched by 13‰ (e.g., Farquhar *et al.* 1989; Boutton 1991; Cerling *et al.* 1993; Quade *et al.* 1992). (Slightly modified from MacFadden and Cerling 1996, reproduced with permission of the Society of Vertebrate Paleontology.)

herbivores (hypsodonty indices [HI] <1) with relatively negative (*c.* <−10‰) $\delta^{13}C$ values for their tooth enamel are principally C_3 browsers. Proceeding clockwise, high-crowned herbivores [HI >1] with relatively negative $\delta^{13}C$ values are C_3 grazers. Although this feeding ecology is relatively rare in present-day ecosystems, and confined to regions with cool growing seasons (high elevations, high latitude, or Mediterranean-type climate), evidence will be presented below to show that this adaptation was very common during the middle Tertiary before 7 million years ago. In the upper right-hand quadrant of Figure 8.4, high-crowned (HI ranging from 1 to 3) herbivores with relatively positive $\delta^{13}C$ values (>−2‰) are C_4 grazers. This feeding ecology is very common in modern ecosystems. In the lower right-hand corner, short-crowned herbivores with relatively positive $\delta^{13}C$ values are very rare in modern ecosystems because C_4 browse is exceedingly rare. (In this case, it also is possible

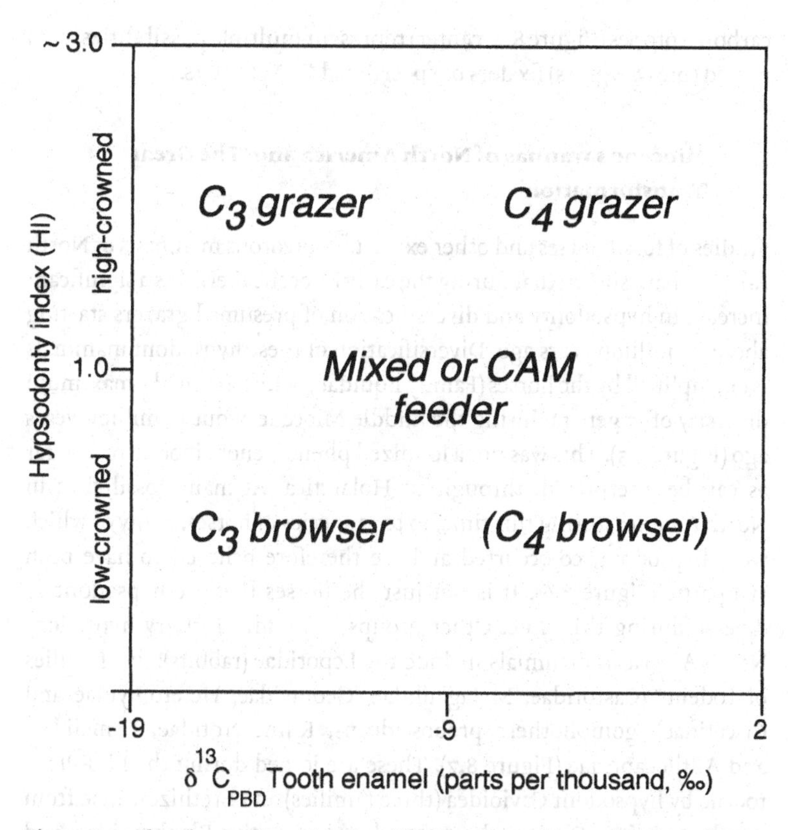

Figure 8.4. Plot of carbon isotopic values ($\delta^{13}C$) versus tooth crown height to show how these two data sets can be used to discriminate C_3 grazer, C_4 grazer, mixed and/or CAM feeder and C_3 browser. (Although a C_4 browser is theoretically possible in this matrix, C_4 browse is very rare in nature. Given the intermediate and wide-ranging $\delta^{13}C$ values for CAM plants, this matrix cannot discriminate between a specialized succulent browser versus a generalized, mixed feeder.) Hypsodonty index is the ratio of unworn/little-worn molar crown height to the anteroposterior molar length. Low-crowned (brachydont) herbivores have a tooth in which the HI is <1; high-crowned (hypsodont or hypselodont) herbivores have a HI that is >1. Maximum observed HIs are usually <3. (From MacFadden and Shockey 1997, reproduced with permission of The Paleontological Society.)

that the particular species in question was feeding on CAM plants.) This category also could include a C_4 grazer which is the exception to the general rule; thus the hippopotamus (*Hippopotamus amphibius*) is short-crowned, but primarily a grazer (Owen-Smith 1988) with relatively positive $\delta^{13}C$ values (-4.2‰, $N=7$, from Amboseli Park, Kenya; Bocherens *et al.* 1996). Species with intermediate values for both crown height and carbon isotopes (Figure 8.4, center) represent multiple possibilities, e.g., mixed (browse/grass) feeders or specialized CAM feeders.

Miocene savannas of North America and 'The Great Transformation'

Studies of fossil horses and other extinct herbivorous mammals of North America have shown that during the early Miocene there was a significant increase in hypsodonty and diversification of presumed grazers starting about 20 million years ago. Diversification of these hypsodont mammals is exemplified by the horses (Family Equidae), which reached a maximum diversity of 13 genera during the middle Miocene about 15 million years ago (Figure 8.5). This was not a localized phenomenon; it occurred, so far as can be interpreted, throughout Holarctica. At many fossil sites in North America during this time, 10 to 12 species of horses, many of which were hypsodont, co-occurred and are therefore believed to have been sympatric (Figure 8.6). It is not just the horses that are hypsodont or diverse during this time. Other groups of middle Tertiary hypsodont North American mammals include the Leporidae (rabbits), five families of rodents (Castoridae, Mylagaulidae, Geomyidae, Heteromyidae and Cricetidae), gomphothere proboscideans, Rhinocerotidae, Camelidae, and Antilocapridae (Figure 8.7). These are joined during the Plio-Pleistocene by hypsodont Cavioidea (three families) and Erethizontidae from South America. During the Late Pleistocene, the Elephantidae and Bovidae dispersed into North America from the Old World. Thus the dental structure and presumed feeding ecology of Late Tertiary herbivorous mammals from North America underwent a dramatic change that Simpson (1951), in referring to horses, called 'The Great Transformation.'

The explanation for this major shift in feeding adaptation has been fairly well accepted by paleontologists since the late nineteenth century. Although the oldest fossil grasses in North America are known from the Eocene of Tennessee (Crepet and Feldman 1991), thereafter, during the middle Tertiary, grasses or their associated remains (e.g., phytoliths) are rarely encountered as fossils until the Miocene. Accordingly, it has been speculated that grasslands began to spread as dominant biomes during the early Miocene and opened up a significant new food resource to be exploited by herbivores. As mentioned above, because of the highly abrasive nature of grasses, the evolutionary cost to potential grazers was the acquisition of high-crowned teeth. This opportunity was exploited by

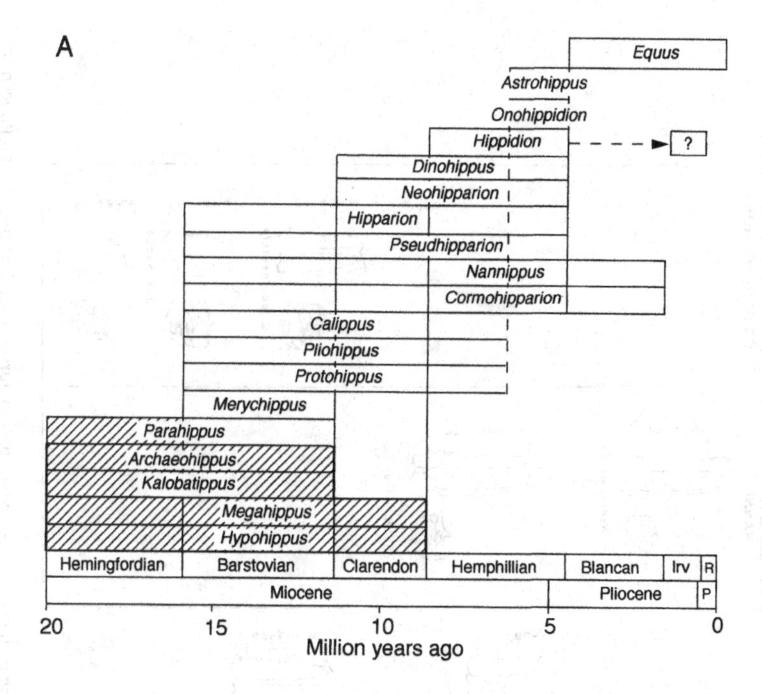

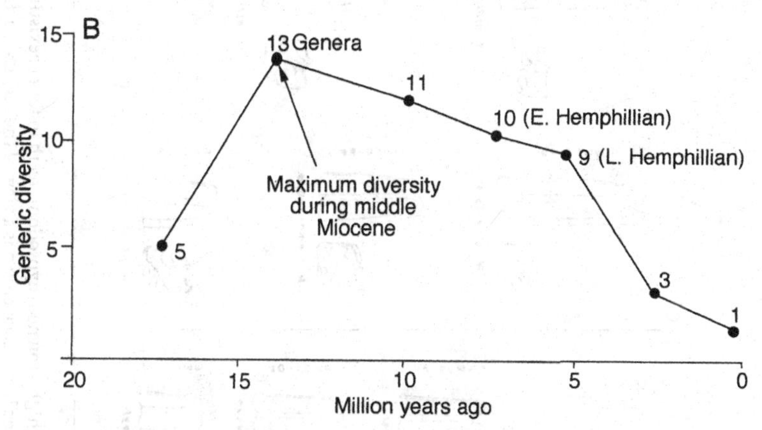

Figure 8.5. (A) Chart showing diversity of presumed browsing (shaded) and grazing (unshaded) horses. (B) Overall diversity of fossil horses during the past 18 million years. (From MacFadden 1992, reproduced with permission of Cambridge University Press.)

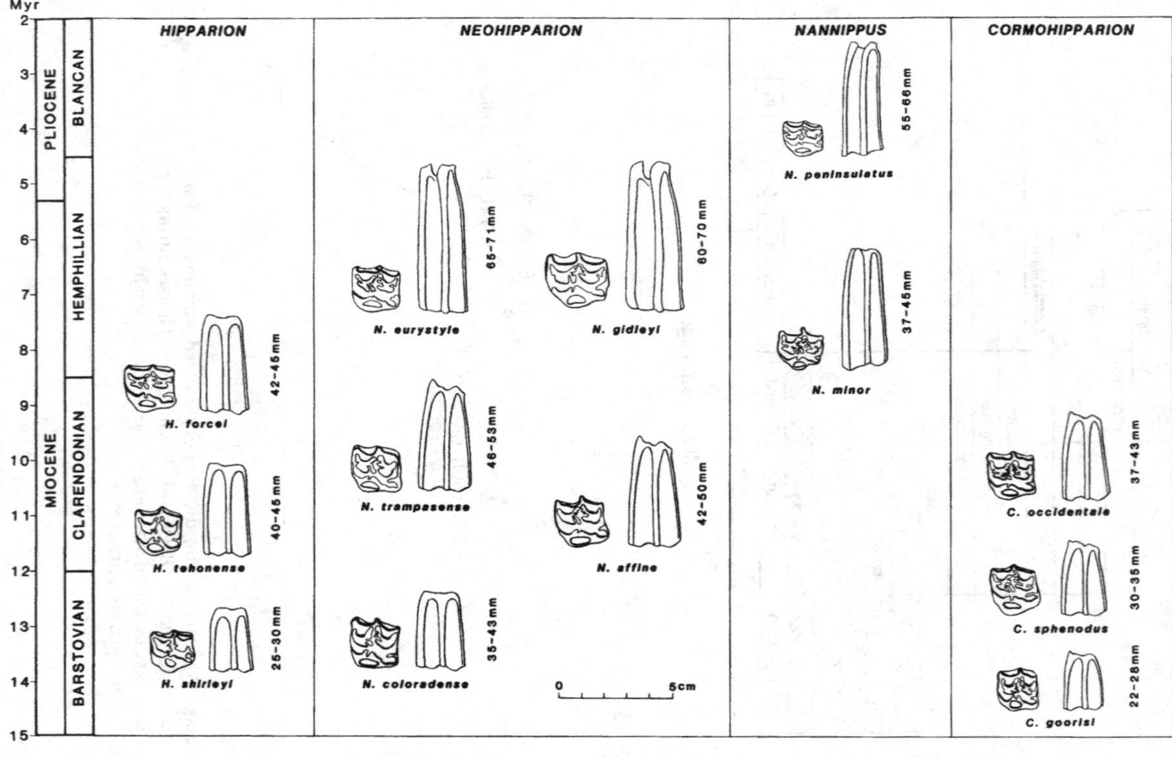

Figure 8.6. Acquisition of high-crowned (hypsodont) teeth in four coexisting genera of three-toed hipparion horses from the Miocene and Pliocene of North America. To the left of each crown is a view of the occlusal surface. Maximum unworn or little worn crown heights are indicated to right of crown. During the middle Miocene (Barstovian) hipparions were hypsodont (hypsodonty indices *c.* 1) and by the Hemphillian they were very hypsodont (hypsodonty indices ranging from 2 to 3). (From MacFadden 1985, reproduced with permission of The Paleontological Society.)

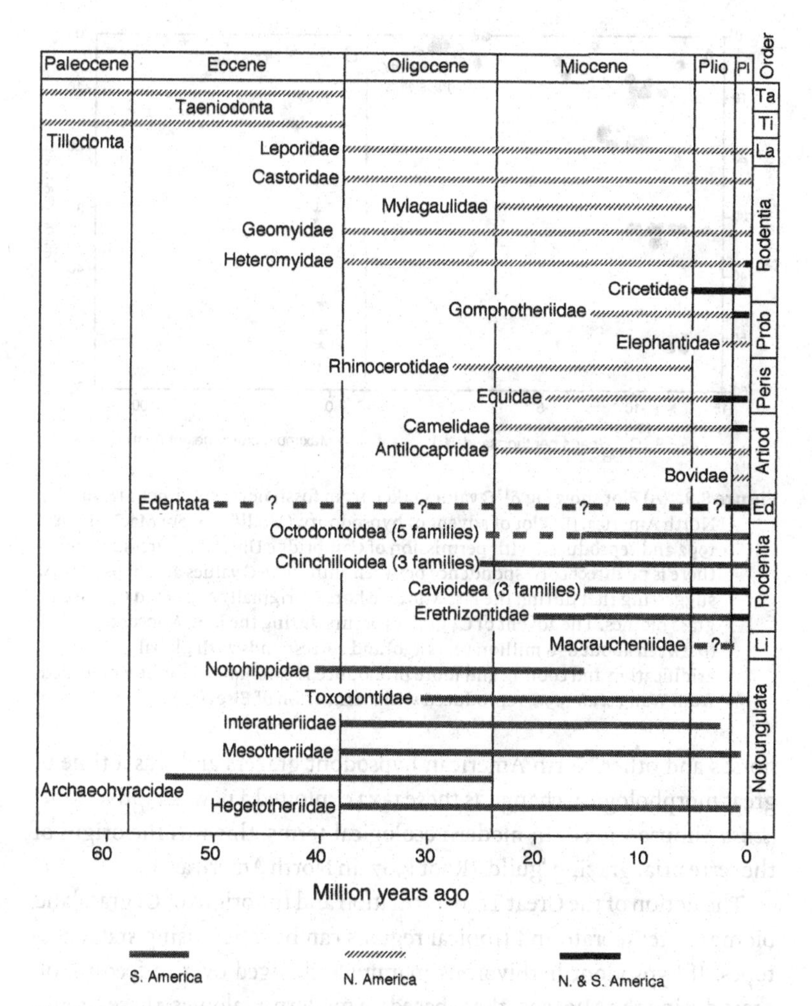

Figure 8.7. Time chart showing the distribution and taxonomic composition of the predominant hypsodont mammals in North and South America. Interpretation of grazing feeding ecology is inferred from either the presence of high-crowned or ever-growing teeth in the clade, or evidence from carbon isotopes, as discussed in the text. The grazing adaptations in edentates and litopterns are questionable. Only the portion of a particular clade's range with high-crowned taxa is included here. For example, fossil horses are known in North America beginning in the Eocene, but only become high-crowned in the middle Miocene, the latter of which is shown here. (Compiled with slight modification from several sources: Patterson and Pascual 1972; Webb 1977; Webb 1978; from MacFadden [1997], reproduced with permission of Elsevier.)

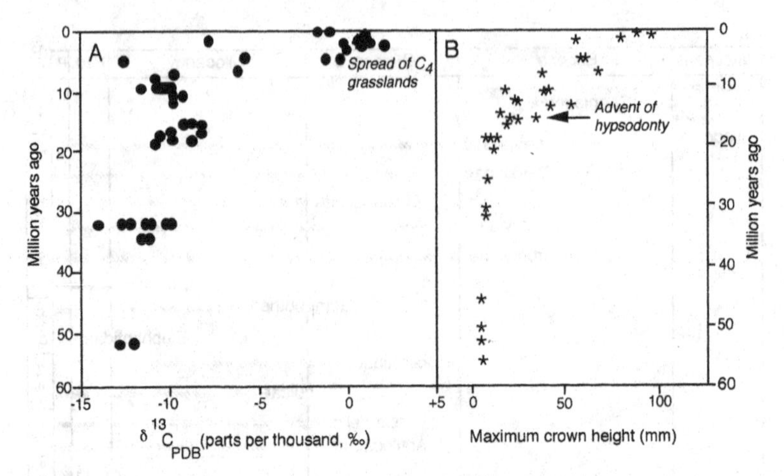

Figure 8.8. (A) Plot showing $\delta^{13}C$ values taken from fossil horse specimens from North America. (B) Plot of advent of hypsodonty. (Modified from MacFadden 1992 and reproduced with permission of Cambridge University Press.) Note there is no direct correspondence between shift in $\delta^{13}C$ values and hypsodonty, suggesting that during the late Miocene horses originally exploited C_3-based grass biomes. The advent of C_4 grasses occurs during the later Miocene (between about 8–6 million years ago) and corresponds with global aridification and cooling and more pronounced seasonality. (Slightly modified from Wang *et al.* 1994, reproduced with permission of Elsevier.)

horses and other North American hypsodont grazers and was a time of great morphological change as these taxa exploited a new 'adaptive zone' (*sensu* Simpson 1953). In modern ecological terms, this was the origin of the terrestrial grazing 'guild' (Root 1967) in North America.

The notion of the Great Transformation and the origin of C_4 grassland biomes in temperate and tropical regions can be tested using stable isotopes. If hypsodont herbivorous mammals changed over to feeding on grass during the Miocene, then, based on modern analogies, there should be a corresponding shift from a C_3 (browsing) to C_4 (grazing) signal recorded in the carbon isotopic values of their tooth enamel. This is exactly the hypothesis that was tested several years ago and the results were very surprising. Wang, Cerling and MacFadden (1994) analyzed the $\delta^{13}C$ values from 40 fossil equid specimens ranging from the Eocene (including *Hyracotherium*, or 'Eohippus') to the Late Pleistocene *Equus*. Not surprisingly, prior to the Miocene, all fossil horses have tooth enamel carbonate $\delta^{13}C$ values of about −10‰ or less (Figure 8.8), indicating a C_3 browsing signal before the advent of grasslands. Surprisingly, in the early and middle Miocene during the time of rapid increase in hypsodonty

(Figure 8.8B) and presumed spread of grasslands, $\delta^{13}C$ values of hypsodont horses remain relatively negative, contrary to what would have been predicted. The anticipated isotopic shift to a dominantly C_4 signal in hypsodont taxa does not appear until about 7 million years (Figure 8.8A). This shift is dramatic, with approximately a 10‰ isotopic enrichment within a relatively short time (i.e., less than one million years) at about 7 million years ago. The explanation for this lag in the advent of hypsodonty and spread of grasslands was initially perplexing. However, it fits into a global model that has developed for world's ecosystems. Carbon isotopic values from a variety of sources, including terrestrial mammals and paleosol carbonates, show that prior to 7 million years ago, the world's ecosystems were fundamentally C_3-based. (There are some fossil C_4 grasses prior to 7 million years ago, but they apparently were neither common nor widespread [Tidwell and Nambudiri 1989].) The advent of C_4-based ecosystems at about 7 million years ago corresponds to the late Miocene 'global carbon shift' (Figure 8.9), including, in terrestrial ecosystems, the beginning of C_4 grasslands as we know them today (Cerling 1992; Cerling *et al.* 1993, 1997). The most probable explanation for the changeover to C_4 grasses after 7 million years ago relates to levels of atmospheric CO_2, more pronounced seasonality, and increased aridity. At high levels of atmospheric CO_2, C_3 photosynthesis is favored, whereas, at low levels of atmospheric CO_2 and high seasonality (including periods of high water stress), C_4 plants are favored. C_4 photosynthesis apparently arose independently numerous times in various plant groups and, from an evolutionary point of view, it is relatively plastic (Ehleringer *et al.* 1991, 1997). Returning to early Miocene hypsodont mammals, the current explanation for the lack of an isotopic shift seems to indicate that these herbivores were grazing, but during this time they fed on C_3 grasses, which were widespread in terrestrial ecosystems prior to 7 million years ago. Evidence in support of this hypothesis comes from other paleontological data. For example, fossil grass parts (anthoecia) have been found in the tooth enamel infoldings of the 10 million year old hypsodont rhinoceros *Teleoceras* from Nebraska (Voorhies and Thomasson 1979; Figure 8.10).

After 7 million years ago, hypsodont mammals shifted to feeding on C_4 grasses that became widespread and formed the origins of the modern grassland biomes. Other than the fact that it seems to have been profound, the overall effect of this shift on the existing grazing guilds is still unclear. It is known that this was the time of dramatic drop in diversity in

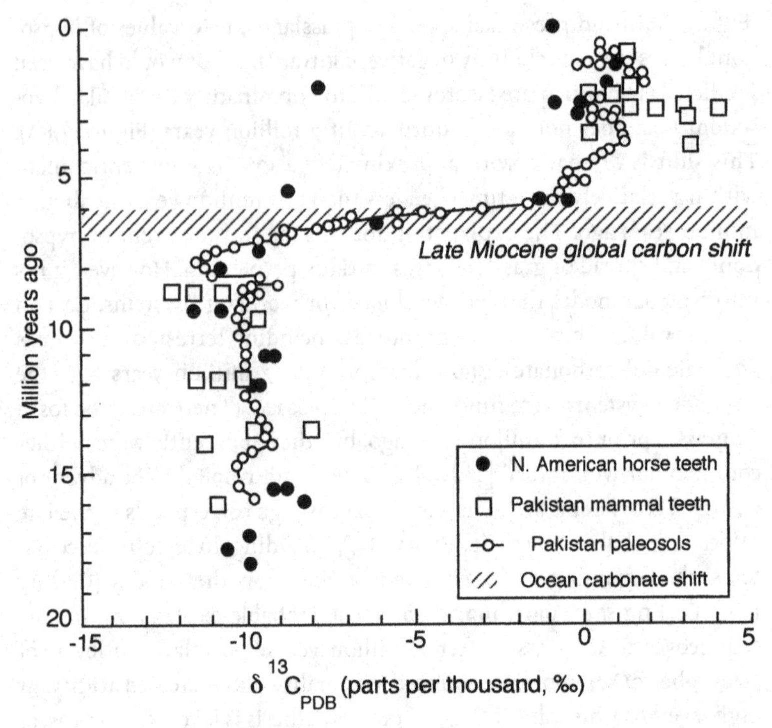

Figure 8.9. $\delta^{13}C$ values of fossil tooth enamel and paleosols from Pakistan and fossil tooth enamel from North America, and the timing (but not same magnitude of $\delta^{13}C$) of the carbon shift preserved in oceanic microfossils during the Miocene. (Slightly modified from Cerling *et al.* 1993, reproduced with permission of Macmillian.)

groups such as horses (Figure 8.5; MacFadden 1992; Cerling *et al.* 1997) and other terrestrial mammals in North America (Webb 1977, 1983).

South American grasslands and 'precocious hypsodonty'

South America drifted during most of the Tertiary as an island continent, and on it evolved a unique fauna of herbivorous mammals. This rich fossil record reveals that there was an abundant and diverse group of high-crowned herbivores, the Order Notoungulata, that evolved elongated, or ever-growing (hypselodont), teeth (Figure 8.11). Based on analogy to taxa from the Northern Hemisphere, this dental adaptation was presumably for grazing. Interestingly, the advent of high-crowned presumed grazers in the South American fossil record predates the origin of ecological

Figure 8.10. Superbly preserved skeletons of the 10-million-year-old hypsodont rhinoceros *Teleoceras* from the Ashfall Fossil Beds of Nebraska. (Reproduced with permission of the Nebraska State Museum, University of Nebraska.)

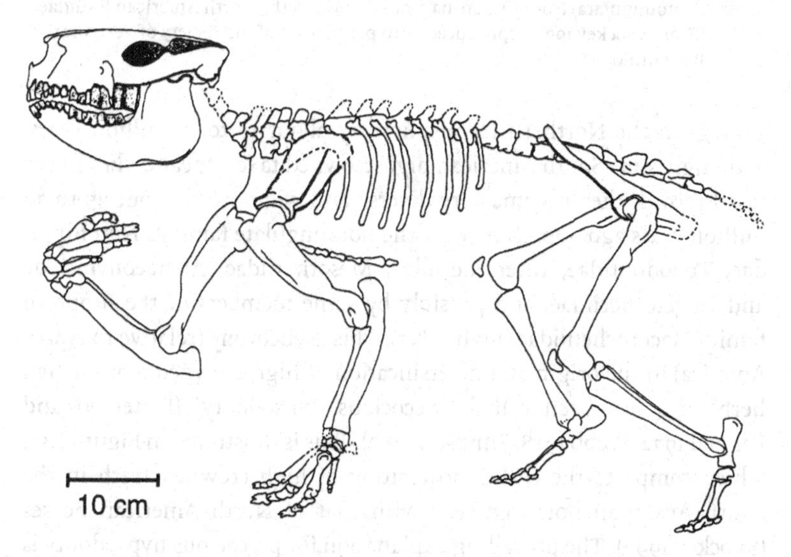

Figure 8.11. *Eurygenium*, an example of a South American notoungulate (Family Notohippidae) with very high-crowned (ever-growing, hypselodont) teeth, from Salla, a late Oligocene/early Miocene locality in the Andean highlands of Bolivia. This curious beast appears to have a disproportionately large cranium relative to its body size, and this is in no small way related to the enormous teeth, which have crowns that extend dorsally into the palatal region. (From Shockey 1997, reproduced with permission of the Society of Vertebrate Paleontology.)

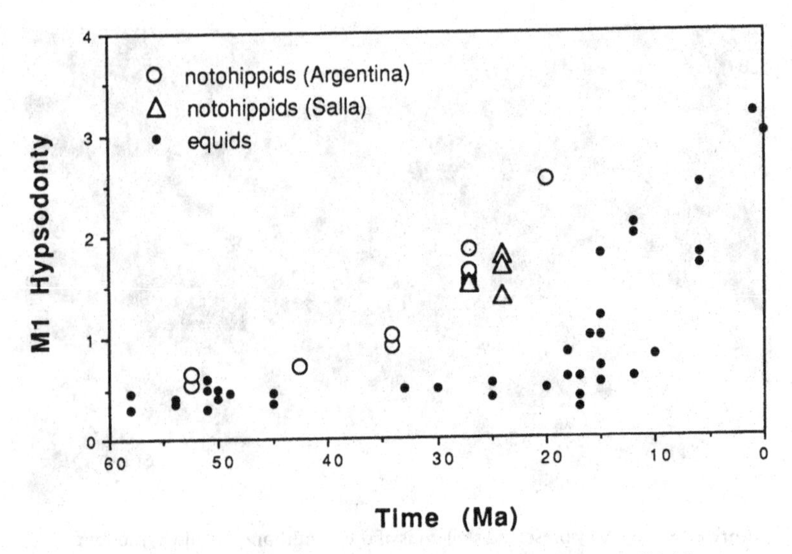

Figure 8.12. Comparisons of the timing of acquisition of high-crowned (either hypsodont or hypselodont) teeth in the South American Notohippidae (Notoungulata) from Argentina and Bolivia and the North American Equidae. (From Shockey 1997, reproduced with permission of the Society of Vertebrate Paleontology.)

analogs in the Northern Hemisphere by about 10 to 15 million years. Although some South American high-crowned taxa appear earlier in the fossil record, they become very diverse and widespread about 35 to 30 million years ago, represented by the notoungulate families Notohippidae, Toxodontidae, Interatheriidae, Mesotheriidae, Archaeohyracidae and Hegetotheriidae, and possibly by some members of the litoptern family Macraucheniidae (Figure 8.7). This asynchrony (relative to North America) in the origin and diversification of high-crowned mammalian herbivores has been called 'precocious hypsodonty' (Patterson and Pascual 1972; Webb 1978; Simpson 1980). This is illustrated in Figure 8.12, which compares the earlier acquisition of high crowned teeth in the South American notoungulates with that of North American horses (Shockey 1997). The prevailing explanation for precocious hypsodonty is that grasslands spread earlier (about 30 to 35 million years ago) in South America than they did in North America (e.g., Stebbins 1981). The poor fossil record of grasses from South America currently does not offer any hard evidence to support this idea.

Like the North American counterparts from the Great Transformation, the advent of grasslands and high-crowned teeth in South American herbivorous mammals could also be tested with carbon isotopes. MacFadden *et al.* (1994, 1996) showed that, although many notoungulate families were hypsodont or hypselodont during the middle Tertiary, carbon isotopic results indicate that C3 foodstuffs, presumably grasslands, predominated until 7 million years ago. Thereafter, as in North America and elsewhere in the world (Cerling 1992; Cerling *et al.* 1993, 1997), C4 grasses predominated in low-elevation equatorial and mid-latitudes that sustain grassland biomes. Using paleosol carbonates from a Miocene–Pliocene sedimentary sequence in northern Argentina, LaTorre, Quade and McIntosh (1997) showed the details of the carbon isotopic shift recorded globally elsewhere, thus providing additional evidence of the origin of C4-based terrestrial ecosystems.

Antiquity of C3 and C4 grasses and grazing guild patterns

Studies of hypsodonty and carbon isotopes together reveal a pattern in which the earliest grasslands of the middle Tertiary were fundamentally C3-based. It was under this regime that grazing mammal guilds originated and diversified. When C4 grasses and grassland biomes became common after the late Miocene global carbon shift 7 million years ago, grazers correspondingly shifted to this new type of grass. As exemplified by horses (Figure 8.5) and also other herbivores, a significant drop in grazing diversity during the past 7 million years seems related to the lower productivity of more arid grasslands (Lauenroth 1979). The mix of modern C3 and C4 terrestrial grassland communities as we know them today have existed over the past 7 million years, which, in a geological context, is a very short period of time.

The study of ancient grazing mammal guilds reveals several patterns in common, regardless of when and where these guilds existed:

1. Grazing mammals span a wide range of body sizes, from tiny rodents to large proboscideans. At the larger end of this continuum, it has been suggested that, relative to medium-sized mammalian grazers, larger-bodied grazers are better adapted for digesting forage of poor nutritive value (Schmidt-Nielsen 1975). This relates to factors such as mass-specific metabolic rate and the longer residence time in the gut for enzymes to digest nutrients from the grass cells (Owen-Smith 1988).
2. Within a given guild there usually is a continuum of coexisting species including: (1) mixed feeders with a significant proportion of grass in their diets; (2) pure grazers, which feed predominantly on grass; and

(3) 'hypergrazers,' which are highly specialized grazers. In Late Tertiary ecosystems, most taxa within this guild and across this continuum are hypsodont to very hypsodont and $\delta^{13}C$ values of tooth enamel range from about −6‰ (mixed feeder/grazer) to 0 to +2‰ (hypergrazer). In Plio-Pleistocene terrestrial ecosystems in Florida between about 3.5 and 0.5 million years ago, carbon isotopic evidence indicates that extinct species of *Equus* occupied the hypergrazer niche. Preliminary evidence suggests that after 0.5 million years ago the Old World immigrant *Bison* either displaced *Equus* as the hypergrazer, or co-occupied this specialized feeding adaptation (MacFadden and Cerling 1996). In South America after the late Miocene global carbon shift at 7 million years ago, but before the Great American Interchange 2 to 3 million years ago, notoungulates such as toxodonts seemed to have occupied the hypergrazing niche. They were later displaced by northern immigrants, including *Lama* and *Equus* (MacFadden *et al.* 1996; MacFadden and Shockey 1997).

3. Species within the same ecosystem originating from the same phylogenetic stock will tend to separate or differentiate their diets as one way to minimize competition for limited resources. Although we do not have definite fossil evidence for this process evolving through time, there are some examples that may serve to illustrate the net result of this competition. The exceedingly rich middle Pleistocene (1 to 0.5 million years old) mammal fauna from Bolivia contains two families of herbivores, the Camelidae (Laminae) and Equidae, in which three species each coexisted. Based solely on dentitions, the lamas (*Palaeolama*, *Lama*, and cf. *Vicugna*) have relatively short crowns (all HI = 0.7) and the horses range from hypsodont (*Hippidion* [HI = 1.9] and *Onohippidium* [HI = 2.2]) to very hypsodont (*Equus* [HI = 3.1]). Despite a lack of niche differentiation within each of these families based solely upon the crown heights, tooth carbon isotopic results show that these coexisting species did indeed have different diets. For the lamas, a small data set suggests that *Palaeolama* was a C_3 browser ($\delta^{13}C$ = −11.4‰), whereas cf. *Vicugna* ($\delta^{13}C$ = −4.5‰), and *Lama* ($\delta^{13}C$ = −3.5‰) were primarily mixed feeders/C_4 grazers. Similarly, for the horses *Hippidion* was principally a C_3 browser ($\delta^{13}C$ = −8.8‰), *Onohippidium* was a C_3 browser/mixed feeder ($\delta^{13}C$ = −7.3‰), and *Equus* was closest to a pure C_4 grazer ($\delta^{13}C$ = −3.4‰; MacFadden and Shockey 1997). An interesting question arises from these data: Why did not the hypsodonty alone discriminate the niche separation in feeding ecology within these two groups? The answer to this might relate to the concept of phylogenetic constraint, where once structures (such as high-crowned teeth) evolved in the ancestors of the species in

question, there is no selective advantage to losing this structure if the descendant returns to a browsing diet.

Conclusions and summary

'Grazers' are defined as animals with diets consisting predominantly (>90%) of grass. Mammalian grazers are characterized by the presence of high-crowned (hypsodont or hypselodont) cheek teeth, which appear to be a coevolutionary response to the pervasive presence of highly abrasive silica phytoliths in grasses. Based on the timing of acquisition of high-crowned teeth, the rich fossil record of extinct mammalian herbivores indicates that grazing first developed as a major feeding guild during the Miocene. Although some fossil grasses have been reported earlier in the Cenozoic, it seems that the rapid acquisition of high-crowned teeth in many clades of extinct herbivores was a coevolutionary response to the advent of widespread grassland communities during the Miocene. The grazing guild originated and evolved independently in terrestrial communities in South America (primarily Order Notoungulata) and North America (orders Rodentia, Proboscidea, Artiodactyla and Perissodactyla) prior to the formation of the Panamanian land bridge.

In addition to dental evidence pertaining to grazing, within the past decade stable carbon isotopes have rapidly gained acceptance as another tool to understand ancient herbivore diets, paleoecology, and global change. Using both tooth structure and carbon isotopes, we can discriminate among C_3 browsers and C_4 grazers, both of which are very common in present-day ecosystems, and C_3 grazers, which are rare in modern ecosystems.

This integrative approach using morphology and isotopes suggests that grasslands and the grazing guild probably originated slightly earlier in South America (*c.* 35 to 30 million years ago) than in North America (*c.* 20 to 15 million years ago). The carbon isotopic data indicate that the earliest spread of grasslands occurred with C_3 grasses (unlike the common C_4 grasses of today). The physiological change to predominantly C_4-dominated grassland communities occurred after the late Miocene 'global carbon shift' about 7 million years ago. The decline in grazing diversity after the late Miocene appears related to increased aridity and lower productivity of grasslands relative those prior to the late Miocene. (Primary productivity is proportional to factors such as rainfall [Lauenroth 1979].) Late Pleistocene herbivore extinctions resulted in a major drop in overall diversity within the grazing guild as we know it in modern terrestrial ecosystems. The origin of the grazing adaptation in the middle Tertiary is a very recent event in the history of terrestrial ecosystems where herbivory is known to have existed for 300 million years.

Acknowledgments

This research has been supported by National Science Foundation grants EAR-9506550 and IBN-9528020 and by the National Geographic Society. I thank Christine M. Janis and Paul Koch for their reviews of this manuscript. This is University of Florida Contribution to Paleobiology number 493.

References

Bocherens, H., Koch, P. L. Mariotti, A., Geraads, D., and Jaeger, J.-J. (1996). Isotopic biogeochemistry (^{13}C, ^{18}O) of mammalian enamel from Pleistocene hominid sites. *Palaios* 11:306–318.

Boutton, T. W. (1991). Stable carbon isotope ratios of natural minerals: II. Atmospheric, terrestrial, marine, and freshwater environments. In *Carbon Isotope Techniques*, ed. D. C. Coleman and B. Fry, pp. 173–195. San Diego: Academic Press.

Cerlíng, T. E. (1992). Development of grasslands and savannas in East Africa during the Neogene. *Palaeogeogr. Palaeoclimatol. Palaeoecol.* (*Global Planet. Change Sect.*) 97:241–247.

Cerling, T. E., Harris, J. M., MacFadden, B. J., Leakey, M. G., Quade, J., Eisenmann, V., and Ehleringer, J. R. (1997). Global vegetation change through the Miocene/Pliocene boundary. *Nature* 389:153–158.

Cerling, T. E., Wang, Y., and Quade, J. (1993). Global ecological change in the late Miocene: expansion of C_4 ecosystems. *Nature* 361:344–345.

Crepet, W. L., and Feldman, G. D. (1991). The earliest remains of grasses in the fossil record. *Am. J. Bot.* 78:1010–1014.

DeNiro, M. J. (1987). Stable isotopy and archaeology. *Am. Sci.* 75:182–191.

Dienes, P. (1980). The isotopic composition of reduced organic carbon. In *Handbook of Environmental Isotope Chemistry. 1. The Terrestrial Environment*, ed. P. Fritz and J. C. Fontes, pp. 329–406. Amsterdam: Elsevier.

Dugas, D. P., and Retallack, G. J. (1993). Middle Miocene fossil grasses from Fort Ternan, Kenya. *J. Paleont.* 67:113–128.

Dyer, M.I., Detling, J. K., Coleman, D. C., and Hilbert, D. W. (1982). The role of herbivores in grasslands. In *Grasses and Grasslands: Systematics and Ecology*, ed. J. R. Estes, R. J. Tyrl and J. N. Brunken, pp. 255–295. Norman, OK: University of Oklahoma Press.

Ehleringer, J. R., Cerling, T. E. and Helliker, B. R. (1997). C_4 photosynthesis, atmospheric CO_2, and climate. *Oecologia* 112:285–299.

Ehleringer, J. R., Sage, R. F., Flanagan, L. B., and Pearcy, R. W. (1991). Climate change and the evolution of C_4 photosynthesis. *Trends Ecol. Evol.* 6:95–99.

Farquahar, G. D., Ehleringer, J. R., and Hubick, K. T. (1989). Carbon isotopic discrimination and photosynthesis. *Annu. Rev. Plant Physiol. Plant Mol. Biol.* 40:503–537.

Hillson, S. (1986). *Teeth.* Cambridge and New York: Cambridge University Press.

Janis, C. M. (1988). An estimation of tooth volume and hypsodonty indices in ungulate mammals, and the correlation of these factors with dietary preferences. In *Teeth Revisited: Proceedings of the VIIth International Symposium on Dental Morphology*, ed. D. E. Russell, J.-P. Santoro, and D. Sigogneau-Russell, pp. 367–387. *Mém. Mus. Natn. Hist. Nat, Paris, sér. C*, 53.

Janis, C. M., and Ehrhardt, D. (1988). Correlation of relative muzzle width and relative incisor width with dietary preference in ungulates. *Zool. J. Linn. Soc.* 92:267–284.

Janis, C. M., and Fortelius, M. (1988). On the means whereby mammals achieve increased functional durability of their dentitions, with special reference to limiting factors. *Biol. Rev.* 63:197–230.

Koch, P. L., Zachos, J. C. and Gingerich, P. D. (1992). Correlation between isotope records in marine and continental carbon resevoirs near the Palaeocene/Eocene boundary. *Nature* 358:319–322.

Kovalevsky, W. (1873). Sur l'*Anchitherium aurelianense* Cuv. et sur l'histoire paléontologique des chevaux. (Prèmiere Partie.). *Mém. Acad. Imp. Sci. St.-Pétersbourg, Sér.* 7, 20(5):1–73.

LaTorre, C., Quade, J., and McIntosh, W. (1997). The expansion of C4 grasses and global change in the late Miocene: stable isotope evidence from the Americas. *Earth Planet. Sci. Lett.* 146:83–96.

Lauenroth, W. K. (1979). Grassland primary production: North American grasslands in perspective. In *Perspectives in Grassland Ecology*, ed. N. R. French, pp. 3–21. Berlin and New York: Springer-Verlag.

Lee-Thorp, J., and van der Merwe, N. J. (1987). Carbon isotope analysis of fossil bone apatite. *S. Afr. J. Sci.* 83:712–715.

MacFadden, B. J. (1985). Patterns of phylogeny and rates of evolution in fossil horses: hipparions from the Miocene of North America. *Paleobiology* 11:245–257.

MacFadden, B. J. (1992). *Fossil Horses: Systematics, Paleobiology, and Evolution of the Family Equidae.* Cambridge and New York: Cambridge University Press.

MacFadden, B. J. (1997). Origin and evolution of the grazing guild in New World terrestrial mammals. *Trends Ecol. Evol.* 12:182–187.

MacFadden, B. J., and Cerling, T. E. (1996). Mammalian herbivore communities, ancient feeding ecology, and carbon isotopes: a 10 million-year sequence from the Neogene of Florida. *J. Vert. Paleont.* 16:103–115.

MacFadden, B. J., and Shockey, B. J. (1997). Ancient feeding ecology and niche differentiation of Pleistocene mammalian herbivores from Tarija, Bolivia: morphological and isotopic evidence. *Paleobiology* 23:77–100.

MacFadden, B. J., Cerling, T. E. and Prado, J. (1996). Cenozoic terrestrial ecosystem evolution in Argentina: evidence from carbon isotopes of fossil mammal teeth. *Palaios* 11:319–327.

MacFadden, B. J., Wang, Y., Cerling, T. E., and Anaya, F. (1994). South American fossil mammals and carbon isotopes: a 25 million-year sequence from the Bolivian Andes. *Palaeogeogr. Palaeoclimatol. Palaeoecol.* 107:257–268.

McClellan, G. H., and Kauwenbergh, S. J. V. (1990). Mineralogy of sedimentary apatites. In *Phosphorite Research and Development*, I. ed. A. J. G. Notholt and I. Jarvis, pp. 23–31. London: Geological Society Special Publications.

Newesley, H. (1989). Fossil bone apatite. *Appl. Geochem.* 4:233–245.

Owen-Smith, N. (1988). *Megaherbivores: The Influence of Very Large Body Size on Ecology.* Cambridge and New York: Cambridge University Press.

Patterson, B. and Pascual, R. (1972). The fossil mammal fauna of South America. In *Evolution, Mammals, and Southern Continents*, ed. A. Keast, F. C. Erk, and B. Glass, pp. 247–309. Albany: State University of New York Press.

Quade, J., Cerling, T. E., Barry, J. C., Morgan, M. E., Pilbeam, D. R., Chivas, A. R., Lee-Thorp, J. A., and van der Merwe, N. J. (1992). A 16 million year record of paleodiet using carbon and oxygen isotopes in fossil teeth from Pakistan. *Chem. Geol. (Isotope Geosci. Sect.)* 94:183–192.

Root, R. B. (1967). The niche separation exploitation pattern of the Blue-gray Gnatcatcher. *Ecol. Monogr.* 37:317–350.

Schmidt-Nielsen, K. (1975). *Animal Physiology: Adaptation and Environment.* Cambridge and New York: Cambridge University Press.

Shockey, B. J. (1997). Two new notoungulates (Family Notohippidae) from the Salla Beds of Bolivia (Deseadean: late Oligocene): systematics and functional morphology. *J. Vert. Paleont.* 17:584–599.

Simpson, G. G. (1951). *Horses: The Story of the Horse Family in the Modern World and through Sixty Million Years of History.* Oxford: Oxford University Press.

Simpson, G. G. (1953). *The Major Features of Evolution.* New York: Columbia University Press.

Simpson, G. G. (1980). *Splendid Isolation: The Curious History of South American Mammals.* New Haven: Yale University Press.

Stebbins, G. L. (1981). Coevolution of grasses and herbivores. *Ann. Missouri Bot. Gard.* 68:75–86.

Sues, H.-D., and Reisz, R. R. (1998). Origins and early evolution of herbivory in tetrapods. *Trends Ecol. Evol.* 13:141–145.

Tidwell, W. D., and Nambudiri, E. V. M. (1989). *Tomlinsonia thomassonii*, gen. et sp. nov., a permineralized grass from the Upper Miocene Ricardo Formation, California. *Rev. Palaeobot. Palynol.* 60:165–177.

van der Merwe, N. J. (1982). Carbon isotopes, photosynthesis, and archaeology. *Am. Sci.* 70:596–706.

Van Valen, L. (1960). A functional index of hypsodonty. *Evolution* 14:531–532.

Voorhies, M. R. and Thomasson, J. R. (1979). Fossil grass antothecia within Miocene rhinoceros skeletons: diet of an extinct species. *Science* 206:331–333.

Wang, Y., Cerling, T. E., and MacFadden, B. J. (1994). Fossil horses and carbon isotopes: new evidence for Cenozoic dietary, habitat, and ecosystem changes in North America. *Palaeogeogr., Palaeoclimatol., Palaeoecol.* 107:269–279.

Webb, S. D. (1977). A history of savanna vertebrates in the New World. Part I. North America. *Annu. Rev. Ecol. Syst.* 8:355–380.

Webb, S. D. (1978). A history of savanna vertebrates in the New World. Part II. South America and the Great Interchange. *Annu. Rev. Ecol. Syst.* 9:393–426.

Webb, S. D. (1983). The rise and fall of the late Miocene ungulate fauna of North America. In *Coevolution*, ed. M. H. Nitecki, pp. 267–306. Chicago: University of Chicago Press.

White, T. E. (1959). The endocrine glands and evolution, no. 3: os cementum, hypsodonty, and diet. *Contrib. Mus. Paleontol., Univ. Michigan* 13:211–265.

Taxonomic index

Subject index